I
A

Industrial Archaeology

A SERIES EDITED BY
L. T. C. ROLT

8

Coalmining

BY THE SAME AUTHOR

The Miners of Nottinghamshire 1881–1914

The Miners of Nottinghamshire 1914–1944

Mining in the East Midlands, 1550–1947

Coalmining

A. R. Griffin, B.A., Ph.D., M.B.I.M.

Longman

LONGMAN GROUP LIMITED
London
Associated companies, branches and representatives throughout the world

First published 1971

ISBN 0 582 12790 4

Printed in Great Britain
by W & J Mackay & Co Ltd, Chatham

Contents

List of Illustrations

Line Drawings in the Text

Acknowledgements

The author wishes to acknowledge permission to use copyright material given by the following: The Historic Society of Lancashire and Cheshire (per Miss P. R. Pleasance) for permission to quote from Mr Anderson's article 'Blundell's Collieries: Technical Developments 1776–1970'; D. Wilcock of the North Industrial Archaeology Society for permission to reproduce his Society's recording sheet; Anderson Boyes Ltd (per R. H. Thorpe) for permission to quote from their company's official history *Fifty Years of Machine Mining Progress* and for plates 22 and 23. Thanks are also due to: Ivor J. Brown for plates 30 and 31; D. E. Bick for plate 8; The National Coal Board (per P. D. Binns, Regional Public Relations Officer) for figures 14 and 17, plates 5, 6, 9–12, 15–17, 24–29, including Nos 15 and 16 taken by the late Rev. F. W. Cobb, former Vicar of Eastwood, which are now deposited with the Board, and Nos 26 and 27 now in the possession of A. Parsons Esq.; the author's son, I. K. Griffin, for figures 1, 2, 5–9, 11, 12, 15, 16, 18, 19 and R. W. Storer for figures 3, 4, 10, 13.

Introduction

Technical history is virtually ignored by many economic historians. One possible reason for this is that books dealing with technical developments tend to be written in involved language intelligible only to an engineer. On the other hand where technical subjects are dealt with in general economic histories, they are often accorded cursory, if not trivial, or downright misleading, treatment.

This series of books on Industrial Archaeology is designed to provide reliable technical history in a form intelligible to the non-engineer. The present volume makes no pretence to being an introduction to mining engineering. The chapters on shafts and winding, and ventilation in particular leave a great deal unsaid. A student of mining engineering may find the book interesting; but he will need to go to textbooks for his mining science.

It may be objected that a technical history for the layman should not be published as industrial archaeology. But what is industrial archaeology?

It is not, as Professor Pollard suggests, 'the study of visible remains for the purpose of illuminating historical knowledge'. That definition would exclude pictorial evidence of things which no longer exist. Further, I do not accept that industrial archaeology merely *illuminates* historical knowledge derived from more orthodox sources. It is itself a source of historical knowledge.

Industrial archaeology is concerned with things rather than documents. By things I mean not only buildings, holes in the ground, sites and artifacts of various kinds; but also people at work and industrial processes. Many interesting things which no longer exist may still be studied because they have been sketched or photographed or recorded in some other way. For example, twelve photographs taken underground at Clifton Colliery in 1895 (a few of which appear in this volume) illustrate a system of mining as it was widely practised between about 1850 and 1920 far more effectively than literary evidence. Such pictures are industrial archaeological evidence.

With mining, some evidence disappears almost as soon as it is unearthed. This applies in particular to old workings exposed on opencast coal sites. The contractors are rarely able to preserve this evidence for more than an hour or two. All one can do is to make as good a record as time will allow; and even that depends on the goodwill of the people on the site.

If industrial archaeologists were to concern themselves only with finding and recording evidence, they would turn their subject into a barren antiquarianism. They need to ask what their evidence means and where it fits in the general pattern.

One of the lessons to be drawn from the present study is that old equipment, techniques and forms of economic organization commonly thought to be extinct often survive here and there. For example, wooden rails which were replaced in most areas by iron plates during the period of the Industrial Revolution were still being used in some small Shropshire coalmines in the 1930s. Similarly, the 'guss', human harness used to draw coal along underground roadways, was still being used in Somerset in the early years of the twentieth century. Again, the 'post-and-stall' pattern of coal working practised in Wales in the seventeenth century has its modern counterpart in small licensed mines in Derbyshire. The equipment used is somewhat different, of course, but the basic method is much the same.

This is true also of the small drift mines operating in the Forest of Dean at the present day. They have all the essential features of early mining practice in that district, and some use equipment like windlasses, which have long since disappeared elsewhere.

It seems, then, that the small unit of production using equipment and methods which are altogether exceptional today, may provide valuable pointers to the typical enterprise of a previous age. The industrial archaeologist must always be on the lookout for late survivals of outmoded equipment and practices. Things in use are more valuable as evidence than artifacts uncovered on a dead site. And the best method of recording old processes is the moving picture. Thus, a film of a Somerset winding engine made by the N.C.B. film unit while the engine was still in use is far more useful than the same winding engine now immured in Bristol Museum. This is not to deny the importance of preserving things in museums when they cannot be adequately preserved *in situ*. There is nothing sadder than to see an industrial monument (by which I mean an industrial building with associated equipment like an engine house and engine) falling into decay. It is far better to preserve things in a museum than not to preserve them at all.

But there are museums and museums. All industrial archaeologists must welcome the establishment of 'living museums', if I may borrow that term from the Ironbridge Gorge Museum Trust. To make the point, one may contrast the working water mill at York Castle Museum with the lace machines in glass cases at Nottingham Castle Museum. For the mining industry there have been several interesting developments recently and reference to these is made in the Gazetteer which follows the main narrative.

Museums can display buildings, machinery and equipment of various kinds but they cannot preserve mine shafts or underground workings. To some extent, the deficiency may be made good by the imaginative construction and display of models; but this is no more than a second best. In the case of the lead mining industry of Derbyshire, there are quite a number of old drift mines, some of them dating back to Roman times, which visitors can enter. To preserve old workings in coal would be far more difficult, but given the will it should be possible to preserve some. For example, where opencast workings strike an extensive series of bell pits, it would be possible to leave one small corner of the field to be excavated carefully by hand, and then preserved. With deep mines, or even drifts, ventilation would present problems, but these need not be insuperable.

But inevitably much of the mining industrial archaeologist's material lies buried, and he must do the best he can with what is available.

In writing this book, I have been helped by many people and it would be impossible to mention them all. There are some (like the old Shropshire collier who used wooden rails in the 1930s) whose names I never knew; there are others whose names I have forgotten. To them, I apologize. I hope that in any case they will accept my thanks.

Among those who have been especially helpful I would like to mention the following:

My colleagues R. Storer (Assistant Area Training Officer), A. J. Williams (Area Chief Engineer) and G. Batey of the North Nottinghamshire Area of the N.C.B.; P. D. Binns, Regional Public Relations Officer; G. Jago, J. Sheldon, G. Walton and W. Hyder of the Opencast Executive; I. Scattergood, late of the North Derbyshire Area; R. Ditchfield, Manager of the South Yorkshire Mines Drainage Unit; F. P. Hillier, who works for the National Coal Board in Somerset; I. J. Brown, who, besides communicating valuable information on Shropshire and elsewhere, also read the manuscript and made many suggestions for its improvement; R. Keddie, who also read the manuscript; B. Trinder, Shropshire Adult Education tutor and an adviser to the Ironbridge Gorge Museum, who introduced me to the Shropshire area; D. E. Bick, who lent me a photograph of the Old Glyn Engine House; J. B. Smethurst, who supplied some information on Lancashire; John Butt of the University of Strathclyde and W. H. Challoner of the University of Manchester; R. F. Youell of Leeds University for information on the Middleton Colliery Railway; my sons C. P. Griffin for information on Leicestershire and I. K. Griffin, who drew some of the sketches; Robin Atthill, who has allowed me to draw on his extensive knowledge of the Somerset coalfield; A. Parsons, who was responsible for preserving the Teversal Colliery photographs; J. B. Plummer, an ex-colleague who now lives in Durham; C. Humphreys for information on Cumberland; Mrs B. E. Palmer, who spent a great deal of her spare time in deciphering my manuscript; and Mrs M. Shepherd, who also helped with typing.

Various people, including N.C.B. Area Chief Engineers, Surveyors and Estates Managers, who supplied information are mentioned in the main narrative or in references, and I cannot thank them too warmly for their help.

I must also place on record my appreciation to Lord Robens and H. M. Spanton, Area Director of the N.C.B.s North Notts Area, for allowing me to accept the honorary academic appointment which made research for this book possible; and to N. Siddall, the Board's Director General of Production who supplied information on the Koepe winding installation at Bestwood besides giving general encouragement.

A. R. GRIFFIN

Department of Economic History
University of Nottingham

CHAPTER ONE

Early Mining Methods

There is no doubt that coal was worked during the Roman occupation of Britain. From archaeological evidence we know that coal was consumed at widely scattered places in this period: mainly military sites like Hadrian's Wall, but also in native villages throughout south-west England and elsewhere. Coal was also used, apparently, to feed a sacred temple fire in Aquae Sulis (Bath). But there are few references to coal in the writings of classical authors, and Professor Nef has argued from this that it cannot have been of great importance in the economy of Roman Britain. There is no doubt that the comparatively small quantities of coal used were supplied mainly, if not entirely, from outcrop workings.

The importance of water transport is indicated by the fact that traces of coal from Nottinghamshire have been found in the Fen villages which were served by an intricate system of canals and navigable rivers.[1]

There is nothing to show whether coal was consumed in Britain during the so-called Dark Ages, though it seems unlikely that its use died out altogether. Coal is not mentioned in Domesday Book, which indicates at least that it was of little importance as a potential source of Royal revenue. But coal was undoubtedly worked in many places during the twelfth century; and for the thirteenth there is a mass of documentary evidence of the mining, marketing and consumption of coal.

However, coal mines were small scale affairs requiring little capital. Many of them represented an investment of £5 or less. Further, the cost of transport restricted the sale of coal to the actual coalfields, except for collieries which were near to the sea or to navigable waterways. The coastal collieries of Northumberland and Durham early developed a virtual monopoly of the London market, and because the coal was transported by sea it was called sea-coal (variously spelt) to distinguish it from charcoal. Similarly, Shropshire owed its development to the River Severn and the South Nottinghamshire field owed its development to the Trent.

Coal was not a popular fuel. For most uses wood or charcoal were preferred. For domestic use, only poor people living in close proximity to the workings burnt coal. Those who could afford wood used it. Except for a few special coals like those mined in the Firth of Forth and south Nottinghamshire areas, the fumes from coal fires were noxious in a period when few places had chimneys, the smoke being allowed to find its way through a hole in the roof. Lime-burners used coal, and so did smiths for forging coarse work, but only where coal was cheaper than the alternative fuel.

Coal did not become really important until the sixteenth century, when timber rose in price much more than other commodities. The production of iron on a fairly large scale, added to the use of wood for a hundred and one traditional purposes, decimated many forests; and Tudor parliaments became concerned for the timber supplies of the Royal Navy.

Various industrial processes—baking, brewing, glass manufacture, and so on—were adapted to the use of coal, either in its raw form or as coke; and the increasing prevalence of chimneys made coal more acceptable as a domestic fuel.

The output of coal expanded rapidly between the middle of the sixteenth century and the end of the seventeenth. There was then a gradual rise, with stagnation in some places, until about 1760. From that date coal (in the form of coke) was increasingly used for the smelting of iron and the consumption of iron rose tremendously. Soon increasing quantities of coal were required for steam engines and for domestic and general industrial use as population expanded. This period of growth of the coal mining industry lasted from about 1760 until about 1920.[2]

The earliest coal workings were on outcrops: coal was picked up where it lay. In the north-east, where seams outcrop on cliffs washed by the sea, coal was gathered on the beaches. By the twelfth century, coal was being got in small quarries and shallow ditches, and in the thirteenth century, in addition to such opencast methods, coal was also being won from shallow drifts and bell pits.

Drifts were found in hilly districts. Indeed, there are still many privately owned drifts being worked under licence today; but whilst these are small by twentieth-century standards they are large by the standards of the thirteenth century. Problems of ventilation, drainage and transport limited the size of the workings. And of course there was plenty of readily accessible coal available without following the seams too far into the hillside.

According to Dr John Butt, many old drift mines can still be seen in Ayrshire, Lanarkshire and East Lothian. One such drift, Avonbraes Colliery, Quarter Hamilton, worked the Ell coal until recently, employing twenty men underground and five on the surface. The coal was hauled by ponies from the workings to a loading platform on the surface. The entrance to the drift is illustrated in Dr Butt's book on the industrial archaeology of Scotland.[3] Most drift mines of about this size are worked by private owners under licence from the National Coal Board, but Avonbraes was worked by the Board itself.

Small drifts are a feature of recent Welsh mining practice, too. Indeed, in the first half of the nineteenth century some Welsh coal works were still in the pre-drift stage of winning coal 'in patches' on the basset edge (i.e. the outcrop of the seam). This involved removing the turf and subsoil covering the shallow seam, one patch at a time, to allow the coal to be worked.

Because so much coal could be won either by the patchwork method or by drifts there was no incentive and little opportunity to master the latest techniques. The few deep mines in existence by the mid-nineteenth century were therefore technologically backward.

Bell-pits are so called because, if they are viewed in section, they have the shape of bells (Fig. 1). A bell pit was sunk like a well shaft down to a shallow coal seam, and the coal at the foot of the shaft was then taken. Next, coal was got from around the pit bottom until the sides were in danger of giving way when the pit was abandoned and another sunk nearby.

These bell pits were usually circular although some were square and others oval. They are of various sizes, though twelve to fifteen feet is perhaps the norm. They were rarely more than about thirty feet deep.

Bell pits are often found during opencast operations. Occasionally, a slice is taken through one, but more often they are seen in plan when the overburden has been removed.

The pits are usually very close together. Occasionally a ditch for drainage was dug through the field first and the pits were then sunk along the line of the ditch. This appears to have been done at Tufty Farm, Yorkshire, which was opencasted recently.

It is rarely possible to see old bell pits on the surface. However, in a field at Stretton, Derbyshire, a bell pit pattern could be seen quite

Fig. 1 Bell pit with windlass

clearly in March 1969 when the snow was beginning to thaw; whilst at Wollaton, Nottinghamshire, an aerial survey in 1955 revealed a similar pattern.[4]

Since March 1969 some of the Stretton bell pits whose existence could only previously be inferred have been exposed during opencast operations for coal and clay (Fig. 2). In one part of the field, exposed in late August 1969, the term 'bell pits' appears at first sight to be somewhat inappropriate. Here the coal lies little more than ten feet down. The pits were sunk close together; and when the soil was removed from the workings, during opencast operations, it was seen that one pit bottom was separated from the next by a very narrow pillar of coal. Indeed in a few cases one pit broke through into the next. Because they were so close together there was little belling of

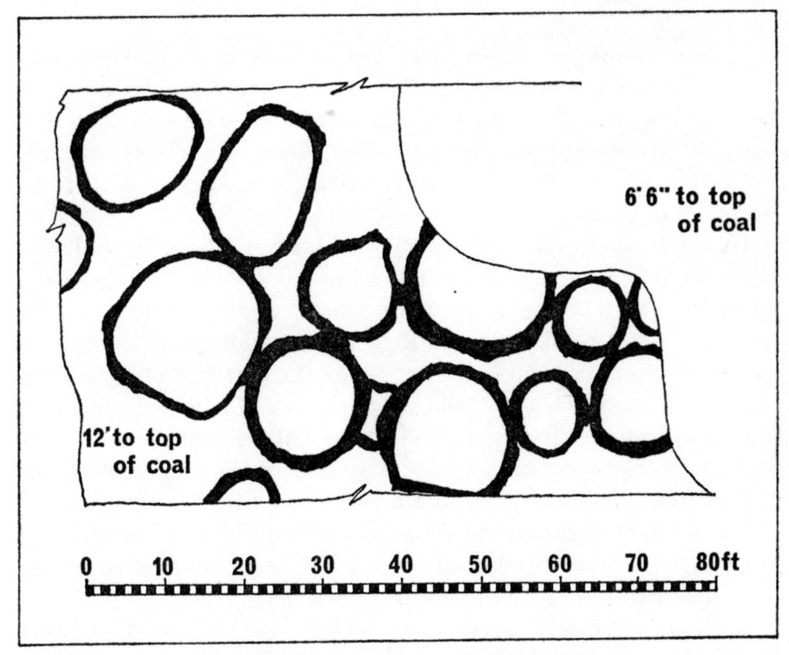

Fig. 2 Bell pits at Stretton, Derbyshire

the sides in the coal seam itself. However, the tops of the pits, judging by the pattern perceived in March, were much further apart than the pit bottoms so that belling out took place during the process of sinking through the soil.

In some parts of Scotland it is possible, according to Dr Butt, to trace the existence of old bell pits because there they were inadequately filled in and surface subsidence has occurred. Similar traces of what are referred to by local people as bell pits are to be found on the Brown Clee Hills in Shropshire, at Wingerworth and Shipley in Derbyshire and at Castercliff, near Nelson, Lancashire.

Disturbed ground on an exposed coalfield is often due to fairly recent working. In every major dispute in the mining industry, particularly in 1893, 1912, 1921 and 1926, many miners won coal from outcrops and took little trouble to fill the shafts properly afterwards. To take only one example, Wessington Common in Derbyshire is in a disturbed condition as a result of outcropping in 1893. It is therefore unwise to assume that all apparent bell pit workings are old.

Where seams lie deeper than twenty feet or so, bell pits are intolerably wasteful of labour. The practice therefore developed of heading out into the seam for a short distance on each side of the shaft. Between the headings the wide pillars of coal were usually strong enough to hold up the roof, although wooden props were no doubt occasionally used.

In some medieval workings entered from Chopwell Colliery, Durham, twenty-odd years ago there were no props or other supports set at all. The headings were driven, about five or six feet wide, in the coal. Cross headings were driven from the main heading at irregular intervals, and a miner who entered these old workings recalls that the whole area had the appearance of a honeycomb. Seen in section, the headings were arch-shaped like a modern tunnel, no doubt to give support to the sides. The apex of the arch touched the rock roof of the seam. The miners who saw these headings called them 'monks' workings' to distinguish them from other old (but much later) bord-and-pillar workings which were sometimes entered by accident from shallow mines in the district.

Arch-shaped headways were not peculiar to Durham. They were found also at Culross in Scotland. Taylor in *The Pennyless Pilgrimage* (1618), quoted by Nef, speaks of a mine with an underground roadway a mile in length under the sea as being 'artificially cut like an arch or vault, all that great length, with many nookes and by-wayes; and it is so made that a man may walke upright in the most places both in and out'. It seems likely that the arch shape was fairly general in this early period.

The headings were restricted in length by the problem of ventilation. This problem was eased, however, where cross headings were driven from one main heading to another, facilitating a flow of air.

This system of heading into the seam, leaving pillars of coal to support the roof, was adopted in all coalfields, though the pattern varied between one district and another. In some districts, as the system developed, the headings became perfectly square in plan and evenly spaced, in others they were irregular. One of the more common variants of this 'stall-and-pillar' system is shown in the accompanying diagram (Fig. 3).

Opencast operations reveal that even in one small area the pattern of working varied. It should be remembered, however, that a large opencast site will cover an area where coal was won piecemeal over a long period of time, possibly two or three hundred years or even more. Thus, on a site near Denby in Derbyshire, there are bell pits, areas where coal was partially extracted by headings, indications of more regular stall-and-pillar work and areas from which the coal was completely extracted.

Complete extraction can mean one of two things. It can mean that the coal has been won by stall-and-pillar advancing from the pit bottom, with the pillars subsequently being extracted whilst retreating to the pit bottom. The first stage in this process was known as working in the 'whole', while the second stage (i.e. extracting the pillars) was known as working in the 'broken'. Alternatively, complete extraction can imply the existence of the more modern longwall system, which we shall study in some detail in a future chapter.

For the moment, a simple explanation of early longwall working will suffice (Fig. 4). Here, a roadway was driven from the pit bottom for a short distance and a coal face was then opened out on either side of the roadway forming a T pattern. All the coal along the face was then extracted, leaving a space called the 'gob' or 'goaf', which was partly packed with debris and small coal, allowing the roof to settle down steadily. The roadway also had to be supported through the gob by dirt packing and by timber.

The longwall system, because it exposed a wide area of roof at a time, necessitated the use of many wooden props (often called 'puncheons') at the coalface. The presence of a large quantity of wooden props in old workings exposed on an opencast site indicates longwall rather than stall-and-pillar.

Shropshire is thought to be the place where longwall originated in the late seventeenth century. From Shropshire the system spread gradually throughout the Midlands, becoming fairly general by the early nineteenth century, though some shallow small mines in the Midlands still employ stall-and-pillar even today.

An interesting example of early (possibly mid-eighteenth-century) longwall is to be found on the Swan Farm opencast site near Wellington in Shropshire, which was still exposed in July 1969. Here there are four seams exposed in the quarry face (the 'high wall' in opencast terminology). In the case of the bottom seam (called the yard seam) the slack and small coal forming perhaps a third or more of the total coal extracted were thrown back into the gob. Because the over-

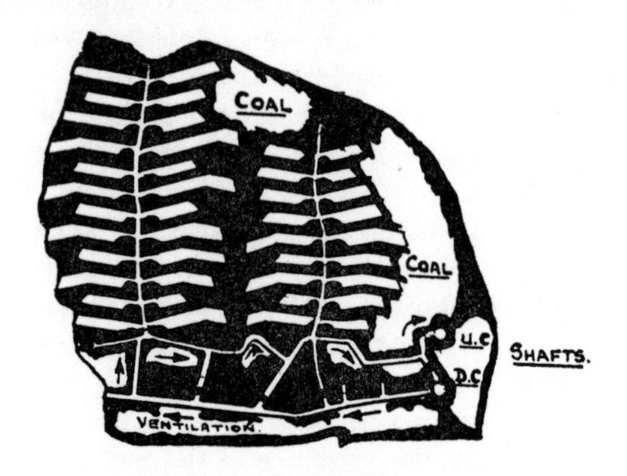

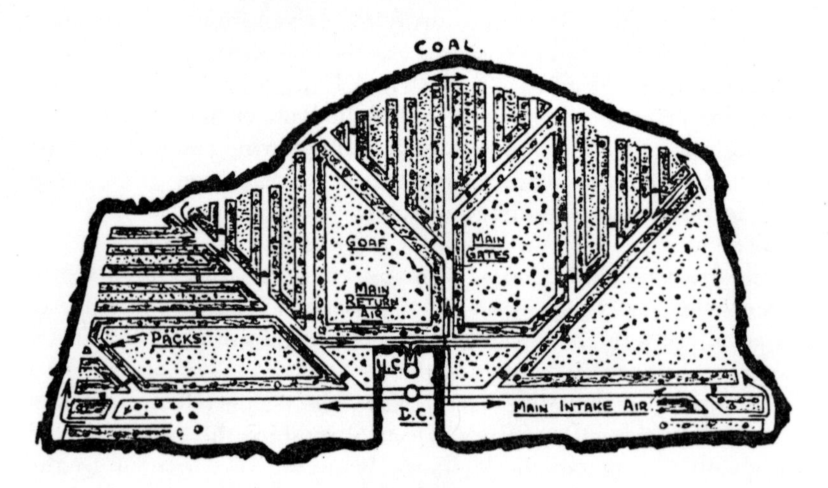

Figs. 3 and 4 Two systems of seventeenth to eighteenth century coalworking. Ventilation note: DC—downcast shaft, UC—upcast shaft, → for direction of airflow

lying strata have settled down on top of the slack, the gob appears at first sight to be a virgin seam of coal.

This practice of throwing all the slack back into the gob was almost universal before the introduction of steam-raising plant which created a demand—albeit a varying one—for slack. A similar example was exposed at Shirland, Derbyshire, in March 1969. In this case an old roadway in the deep soft seam could be seen in the high wall running at right angles to the workings. On one side of the roadway the seam was virgin, some four feet in thickness. On the other side it had been completely extracted but the slack had been stowed in the gob. This side was reduced to about two feet in thickness, the pressure of the overlying strata having compressed the slack to form, as in the Shropshire example, what appeared at first to be solid coal, although it was actually quite loose.

In another part of the Stretton site (away from the bell-pits which we have previously mentioned) where the coal lies at a depth of 40 feet or more, a cut was taken by the contractor's digger through some old workings in August 1969. Here, a roadway supported by wooden props and lids was exposed. The roadway, only a few feet from the pit bottom, was driven in the coal only, and was some 2 feet 6 inches high. A further foot of coal was left up to support the roof, the stratum above the coal being rather friable. The width of the roadway cannot be given, because only one side of it was intact. However, the cut was slightly oblique and it is possible to infer from the coal left around the pit bottom that the roadway was quite narrow. At right-angles to this main roadway was another, so narrow as to deserve the term 'snicket'. On either side of the snicket, and level with its roof, was a line of holes which had evidently been cut into the coal top for horizontal wooden bars to support the main road at the junction. The roof at this point was particularly crumbly. One is tempted to conclude that these were roadways in a mine working longwall. In the absence of evidence of a longwall face, however, it is impossible to be certain of the pattern of working. This may have been a colliery at the more primitive level of heading out into the seam. This supposition is supported by the fact that there were four shafts in line with each other revealed in the high wall within a distance of about 180 yards. On the other hand, one would have expected the headings to be wider if these were for winning coal rather than for transport from longwall faces.

The longwall system, whilst having certain advantages, did not become universal. In Northumberland and Durham in particular, pillar work (called pillar-and-bord in the north-east) continued to be the usual system until well after the nationalization of the industry. The argument usually advanced in favour of the system was that it was necessary to leave pillars to give support where soft coal was being worked in association with a weak roof and a soft floor.[5] To suggest that these conditions were universal in the north-east and unknown elsewhere is clearly untenable. There was no physical impediment to the introduction of longwall into parts of the north-east, as elsewhere, in the nineteenth century. The reluctance to change even where conditions were ideal for longwall was due to the natural conservatism of men and managements.

Bord-and-pillar had been established in the district since the fifteenth century or earlier.[6] By the late eighteenth century, when longwall began to oust bord-and-pillar elsewhere, the latter system had been developed to a fine art in the extensive collieries of the two northern counties. Most mining engineers saw few advantages in changing.

The men were accustomed to working alone or in pairs in their own separate bords. So as to ensure that every man had a fair share of good and bad work, the working places were drawn by lot (the 'cavel') at quarterly or half-yearly intervals. With the longwall system, on the other hand, men worked cooperatively in teams.

Stall-and-pillar is still used in the Midlands, to give support to the roof, in small licensed mines which are shallow by modern standards. It is also used occasionally at large mines when working under heavily built-up areas where subsidence must be kept to a minimum.

The bord-and-pillar system of working was more conducive to the development of a high level of craftsmanship than longwall. Longwall, on the other hand, was more conducive to the building up of a fast tempo of work.

There is thus a tendency for the Durham or Northumberland miner, with his tradition of craftsmanship, to work to higher (in some cases, perhaps, unnecessarily high) standards than his Midlands counterpart; but to do so more slowly. The conservatism which postponed the adoption of longwall in the north-east may be ascribed to the same cause. The system of working gave rise to the tradition of individual craftsmanship as opposed to team effort; and the tradition of individual craftsmanship helped to maintain the system.

CHAPTER TWO

Transport

Coal is a bulky commodity. It is therefore costly to transport it for any great distance. Before canals were built most collieries served a very small area. The exceptions were those situated near the sea or near a navigable river. As we have indicated the Northumberland and Durham coalfield owed its early expansion to the seaborne trade with London, but considerable quantities were also being shipped to various east coast ports (e.g. Yarmouth, King's Lynn) by the seventeenth century, and much was also being exported to western Europe. From London, northern coal found its way up the Thames Valley to dozens of towns and villages.

Similarly, the Scottish and South Wales coalfields benefited by their proximity to the coast, though their expansion was much less rapid. To illustrate this, we may take the figures for coal exports to foreign countries as estimated by Professor Nef for the years 1551–60 and 1591–1600. Of the total amounting to 12,000 tons per annum in the earlier period, Northumberland and Durham are credited with 10,000 tons against 1,000 tons each for Scotland and South Wales (including Bristol). In the later period, the corresponding figures are 30,000, 7,000 and 2,000 tons respectively.[1]

At the end of the seventeenth century the river traffic in coal is estimated to have been somewhere between 200,000 and 300,000 tons a year compared with about a million tons a year carried by sea.[2] The principal navigable rivers in coal mining districts were the Trent, the Tyne, the Wear, the Severn, the Wye, the Humber and the Clyde. On the Trent, the coal had to be carried in flat bottomed barges and the many sand banks required skilful navigation. On the Severn, however, fair-sized sailing boats were used and freight rates were considerably lower. Again, many of the mines of the Coalbrookdale area of Shropshire were close to the river banks. Consequently, in the seventeenth century the towns of the Severn valley had the advantage of cheap coal, and this played an important part in the expansion of places like Gloucester, Worcester and Tewkesbury.

But most collieries were at some distance from either the sea or a navigable river. These 'landsale' collieries were chiefly dependent on a purely local market. Most of them were therefore very small; although a minority, which were close to growing towns, were larger-scale concerns.

Warwickshire coal, it is true, was transported for distances of fifteen miles and more to places like Leicester, over indifferent roads, but such distances were exceptional because Warwickshire had no competition from waterborne coal. For landsale collieries outside Warwickshire, a ten mile radius was about the usual limit of the market area prior to the mid-eighteenth century. For much of the year the roads were impassable except for pack-horses; and in many districts they were, indeed, the most common means of transport for landsale coal. In some places (e.g. Cossall on the Notts–Derbyshire border) the passage of pack-horses was facilitated by narrow paved ways some of which could still be seen a few years ago. Farm carts were used for transporting coal after harvest, however, when the roads were usually reasonably free from mud. Urban coal merchants thus stocked up with coal for the winter, during the months of autumn. In midwinter the pits usually stood idle, partly because of transport difficulties and partly because the workings were difficult to keep dry.[3] We shall have more to say on that in Chapter Six, on drainage.

Britain's first railroad was constructed by Huntingdon Beaumont in 1604 in order to cheapen the cost of transporting coal from Strelley to Wollaton lane end, near Nottingham. The total length of this railroad was two miles or so, and the cost of transport in 1610 was one shilling a rook (about 25 or 30 cwt) since coal which sold at Strelley for 4*s* to 4*s* 6*d* a rook sold at the rails end for 5*s* to 5*s* 6*d*.[4] These rails were undoubtedly wooden ones, fastened to wooden rails by tree-nails. Later wooden rails were rectangular in section, the flange being on the wagon wheels; and there is no reason to doubt that this was the case at Wollaton. Certainly, special 'wagens' were used on this railway and not ordinary coal 'cariadges'.[5]

A similar railway was constructed at Calcotts, Broseley in Shropshire in 1605 by Willcox, a local colliery proprietor. The wagons were drawn by oxen to the banks of the Severn, a distance of about a mile. Another railroad built by Clifford, one of Willcox's rivals in the Calcotts district, drew coal from pits much closer to the Severn down a steep incline. Here the wagons were let down with a rope. From the evidence it is clear that the Broseley rails were rectangular, the flange being on the wheel; and flanged wheels cut from solid blocks of wood have been found at old workings in the Broseley area. One such wheel found at the Caughley mines, is preserved at the Shrewsbury Public Library and Museum.[6]

From Wollaton, Huntingdon Beaumont went to Northumberland in 1605, and introduced various technical novelties there including his rails. He built wooden railways at Bedlington, Cowpen and Bebside. As we have seen, the collieries of the north-east were favoured by their proximity to the sea, but in many cases the coal still needed to be transported overland for some miles. Thousands of carts were engaged in this process: one large colliery alone was said to have employed 700 carts and wains in the seventeenth century. The introduction of railways lowered these transport costs considerably, besides facilitating the transport of coal in the winter months. Roger North, in 1676, explained that:

> The manner of the carriage is by laying rails of timber, from the colliery down to the river, exactly straight and parallel, and bulky carts are made with four rowlets fitting these rails; whereby the carriage is so easy that one horse will draw down four or five chaldron of coals, and is an immense benefit to coal merchants.[7]

A later writer, Dr Stukely, observed in 1725 that many of the railways were so constructed that the wagons ran down to the river by gravity. He also referred to the recently completed Tanfield (or Causey) Arch built for the railway which carried coals from Tanfield Colliery to the Tyne, a distance of five miles. This is generally accepted as Britain's earliest railway bridge. This arch, which had a span of 103 feet and a height of 63 feet, cost £1,200. The original structure fell down shortly after completion, but was replaced by the present structure, completed in 1727. The architect, fearing a second collapse, is said to have killed himself by jumping from the top of the arch.[8] This arch was described in 1898 as 'a picturesque ruin' but it is now preserved as an ancient monument.

Of course, whilst the wagons ran down to the river by gravity, they needed to be hauled back to the pits when empty and horses were used for this. Very often, a low carriage was attached at the rear of a train of loaded wagons so that the horse could ride one way; an arrangement which the horses are said to have greatly appreciated.[9]

Turning to the South Wales coalfield, Sir Humphrey Mackworth, in the last decade of the seventeenth century, harnessed wind power to drive wagons from his pits near Neath, to the waterside. The trains of wagons were fitted with small sails when there was a favourable wind, and were drawn by horses at other times.

In defending his railroad in a law suit in 1706 or thereabouts, Sir Humphrey said that such 'waggon-ways' were very common in Shropshire, and that their usefulness in preserving the roads, 'which would be otherwise made very bad and deep by the carriage of coal in common waggons or carts', was generally recognized.

Another early wagon-way, at Alloa, Scotland, which used wooden rails at first, went over to iron in 1785. The tunnels built for this line are still preserved.

Shropshire must take the credit for the first use of rails for underground transport. Many of the coal works near the Severn were drift mines. Referring to the Shifnall and Madeley areas, the Rev. Francis Brokesby records the following in 1711:

> Where . . . are considerable coal mines into which they descend not, as in other places, into pits; but go in at the side of a hill, into which are long passages, both strait forward, and from thence on each side; from whence they have dug coles: which, by small carriages, with four wheels of above a foot diameter, thrust by men, they convey not only out of the long underground passages, but even to the boats which lye in the Severn ready to receive them: a sight with which, some years ago, I was not a little pleased.[10]

One such 'long passage', of rather later date (1787), may still be seen. This is the Coalport Tar Tunnel, brick lined, driven into the hillside on the banks of the Severn, which was originally designed to connect with the coal workings but was subsequently used for transporting bitumen, which occurs naturally in that part of Shropshire. Near the end of the tar tunnel was an inclined plane, still discernible, to lift the loads up to the Shropshire canal, completed in 1792. This was a tub-boat canal and the boat's cargoes had to be transshipped into Severn barges some 1,200 yards or so down river.

There is no doubt that Shropshire was technologically advanced in the seventeenth and eighteenth centuries. But the nineteenth century saw a process of ossification. Thus, until well into the twentieth century, some small Shropshire collieries still used wooden rails for underground transport, though elsewhere they had been largely abandoned in favour of iron plates or rails in the late eighteenth or early nineteenth century. One such wooden rail, exposed at Mossey Bank opencast site in July 1969 is $49\frac{1}{2}$ inches long and $1\frac{3}{8}$ by $1\frac{1}{4}$ inches in section. It is made of oak and was spiked direct on to the sleepers by iron dog nails. An old collier working on a nearby site at Little Wenlock recalls using such rails in small mines as recently as 1933. The rails were produced in large quantities in a local wood yard. In earlier practice, wider deal rails were common.

Elsewhere the superiority of iron rails was quickly recognized. The two wooden surface wagon-ways which were still in use in the Derbyshire coalfield in 1808 were reported by Farey as curiosities. At first it was usual to lay plates of iron on top of wooden rails both for surface and underground wagon-ways but, as iron became cheaper, cast iron plates gradually took the place of wooden and composite rails for both surface and underground transport.

A private branch from Oakthorpe Colliery to the Ashby-de-la-Zouche Canal was originally laid down with wooden rails but these were soon replaced by cast iron.

The Middleton Colliery wagon-way, Leeds, perhaps the earliest railroad to be built under the authority of a Private Act of Parliament, also had wooden rails when first constructed in 1758, but iron rails were substituted later. This is the oldest colliery railway still *in situ* and is now preserved, partly under covenants given to the National Trust by Clayton, Son and Co. Ltd in 1962. This railway used a steam-driven locomotive as early as 1812. This was of Blenkinsop's design, which worked on the rack rail and pinion system similar to that used on mountain railways today. John Blenkinsop, the inventor of the rack rail, was the Viewer of the Middleton Colliery. His locomotives were built by the Leeds firm Fenton, Murray and Wood for £400 each; and it was at the suggestion of Matthew Murray that each locomotive was given two double-acting cylinders. This greatly improved design gave the engine a smooth action without the need for a flywheel. The rack-rail enabled these light locomotives to draw heavy weights uphill.[11] The railway is now in the care of the Middleton Railway Trust, an enthusiastic preservation body which maintains both passenger and goods services on the line. Some of the original stone sleepers used in 1812 are exhibited as are some other nineteenth-century relics.

According to Galloway, the earliest public railway was constructed in 1789 by W. Jessop to carry coal limestone and other traffic from the small Leicestershire coalfield (near Ashby-de-la-Zouche) to Loughborough. However, the line to which he refers—the Forest Line of the Leicester Navigation Company—was not authorized by Act of Parliament until 1791 and was not officially opened until 1794. There is some doubt, too, whether it was legally a public railway. It was nevertheless important as probably the first railway to use fish-bellied rails. Before this date cast iron edge-rails had apparently been no more than iron bars, rectangular in section; and the rails used by Jessop on this line have therefore been regarded by some historians as the earliest true iron edge-rails. When the Stephensons built their historic Swannington to Leicester Railway in the same district in 1833 they used rails of the same pattern.

The Forest line was interesting also in being part railway and part canal. At the coalfield end railways ran from Coleorton Common and Swannington Common to join the canal at Thringstone bridge. To the north another railway branch connected the lime quarries of Cloud Hill and Barrow Hill to an arm of the canal. At the other end of the line a railway ran from Loughborough to connect up with the canal at Nanpanton.

Coal for the Leicester market had therefore to travel in horse-drawn wagons by rail for up to a mile and a quarter, then to be transhipped on to the canal to be carried a further five miles as the crow flies (and half as far again as the canal twists) then to be transhipped again to rail for another two miles or more, and finally to be transhipped once again at Loughborough wharf for the final part of its journey on the Leicester Navigation.

Some writers believe that transport costs on the Forest Line were therefore much higher than for Derbyshire coal and that this accounts for the closure of collieries near the canal, which were supposedly unable to compete on the Leicester market. This view has been convincingly challenged in a recent thesis by C. P. Griffin, who points out that the closure of collieries at Coleorton, Swannington and Thringstone was due to other factors.[12]

Because of these colliery closures the Forest Line was very little used and the rails and sleepers were taken up in 1824. These rails were, of course, cast iron, and they were supplied by Butlers of Chesterfield in three-foot lengths. According to the *Repertory of Arts and Manufactures* for 1800, three feet was the usual length of cast iron rails used on colliery railways, and each length weighed from twenty to forty pounds. In one case, at Brinsley, where trains of twenty-one two-ton capacity wagons were drawn by horses down a gradual incline (about 1 in 100), the rails were thirty-three pounds per yard, the gauge four feet two inches and the sleepers either stone or wood. The rolled iron fish-bellied rails used by the Stephensons for their Swannington line in 1833 were much more satisfactory than cast iron in every way: they were stronger, smoother, and made in much longer lengths. The Swannington Incline branch was in continuous use until 1948, and the Stephensons' rails were not taken up until four years later. They were photographed *in situ*.

The canal section of the Forest Line can still be traced in many places. At Thringstone, for example, the canal banks are visible from the main Ashby to Loughborough Road. However, the author has not been able to trace with certainty the course of any of the railway sections, although in places this can be inferred from the position of the disused pit shafts.

Jessop's Forest Line railway was designed for horse traction, like most of the early railways. Some used stationary steam engines, however, especially on steep inclines.

Most wagon ways built in this period in the inland coalfields were constructed as feeders to the growing network of canals. Every mining district became criss-crossed with canals and their associated railways. Between 1770 and 1800 Parliament dealt with 113 Bills for the construction of inland navigations, many of which were in or near coalfields. The importance of coal traffic to the canals is not to be wondered at. Coal is a bulky commodity and transport costs, as we have noted, account for a high proportion of its delivered cost. There was therefore every incentive to ensure that coal could travel for the greater part of its journey by water. And everywhere the demand for coal was increasing as Britain's industries expanded.

Most Acts authorizing the construction of canals made provision for railways to be built in connection with them. As we have seen, early railways used wooden rails but cast iron was coming into use in the later part of the eighteenth century. Unshaped edge-rails were apparently being manufactured at Coalbrookdale as early as 1760; but most of the early cast iron wagon ways used flanged plates, and not edge-rails. Indeed, in the Coalbrookdale district itself, plateways using horse traction were in use early in the present century and their tracks can still be seen in many places. A small wagon, doubtless used for carrying coal, whose plain wheels show that it was used on the nearby plateway, was recovered from the bed of the Shropshire (tub-boat) canal above Blists Hill and is now housed at the Museum there. Another horse-worked plateway surviving into this century was built by Benjamin Outram from Denby to the canal at Little Eaton, Derbyshire, in 1794. It carried coal from Denby until 1908. The line of the tramway can still be traced and one of the wagons and some rail has survived. This tramway was fortunately photographed before closure. The superstructure of each wagon on this line was loose, and at the canal wharf a crane lifted the superstructure with its load of coal and deposited it bodily into a barge. The surviving wagon and some stone sleepers from the Little Eaton line are now kept at the Lound Hall Mining Training Centre. One tramway, using fish-bellied rails, laid down by Jessop and Outram to carry coal from Grantham Canal to Belvoir Castle in 1792, is still in place.

The early steam locomotives ran on plateways. The first is said to be one made by Richard Trevithick which was tried out in 1804 at Penydaren plateway, Merthyr Tydfil. Similarly, *Puffing Billy* was made by Blackett and Hedley for the Wylam Colliery tramway in 1813. There is a delightful drawing of an 'old locomotive engine' on this line in T. H. Hair's *Sketches of the Coal Mines in Northumberland and Durham* (1839), although by that date flanged plates had given way to edge rails. Hair also quotes a notice of this railway in 1825:

> Each engine draws ten waggons that carry eight chaldrons of coals, or 21⅓ tons, which is above two tons and one-tenth to each waggon. Sometimes a dozen or more waggons are dragged by one engine. A stranger is struck with surprise and astonishment on

> seeing a locomotive engine moving majestically along the road, at the rate of 4 or 5 miles an hour, drawing along from ten to fourteen loaded waggons; and his surprise is increased on witnessing the extraordinary facility with which the engine is managed. The invention is a noble triumph of science.

This wagon-way carried coal from Wylam to the staithes at Lemington, five miles distant, where it was loaded into seagoing vessels.

After 1820 flanged plates gave way to rolled iron rails (patented by John Birkinshaw in that year), for tracks where steam locomotive engines were to run. Flanged plates might be adequate for wagon-ways employing traction by horses or stationary engines, but they were unsuitable for locomotives.

So far, as we have seen, railways were subsidiary to water transport. In Northumberland and Durham they carried coal down to the staithes for transhipment by sea; and in inland coalfields they carried it to canal wharfs. Stationary engines were more widely used in the north-east than elsewhere and remains of several engine houses formerly used in connection with colliery wagon-ways still exist. The Stanley haulage engine house, near Crook, Durham, is shown in a photograph used by Atkinson, who also mentions other buildings of a like kind which still stand. The last stationary engine operating on a colliery line in the district was dismantled as recently as 1959 after 123 years' continuous use. This was at Warden Law on the Hetton to Sunderland line, and it has been preserved for the Northern Regional Open Air Museum.[13]

Some inland wagon-ways also used stationary steam engines, particularly on steep inclines. This was so with the Swannington incline, for example. A stationary engine built by the Horsley Coal and Iron Company of West Bromwich was installed at the top of the incline, in 1833 to haul the wagons up to the point at which locomotives could take over. This engine could haul four full wagons weighing twenty-seven tons up the incline, and it would start with steam pressure as low as twenty pounds per square inch. There was no brake on the engine until the 1880s when the Board of Trade insisted that a hand brake should be fitted. The engine continued in use until about 1930.

A poetic account of the wagon-ways in the Great Northern coalfield by a Mr Howitt is quoted with approval in Hair's *Sketches*.

> Here and there, you saw careering over the plain, long trains of coal-waggons, without horses, or attendants, or any apparent cause of motion, but their own mad agency. They seemed, indeed, rather driven or dragged by unseen demons, for they were accompanied by the most comical whistlings and warblings, screamings and chucklings, imaginable. When you came up to one of those mad dragon trains, it was then only that you became aware of the mystery of their motion. They ran along railways, and were impelled by stationary engines at a distance, which stood often in valleys quite out of sight. A huge rope running over pulleys raised a little above the ground in the middle of the railway; and these pulleys or rollers, all in busy motion on their axles, made the odd whistlings and warblings that were heard around. In truth, the sight of these rollers twirling, and the great rope running without visible cause, was queer enough. Amid all these uncouth sounds and sights, the voice of the cuckoo and the corn-crake came at intervals to assure me that I was still on the actual earth, and in the heart of spring, and not conjured into some land of insane wheels and machinery possessed by riotous spirits.

Canals and their subsidiary wagon-ways enabled the inland coalfields to compete more effectively with the coastal coalfields. By lowering drastically the cost of transport and therefore the delivered price of coal, they increased its effective demand and thus boosted the expansion of the industry which was already under way. The typical colliery enterprise grew in size as its market area expanded. Of course there were still many purely landsale collieries too far distant from a canal to make use of water transport, and they remained small, relying on a local market, though even they benefited from the lowering of transport costs brought about by the building of turnpike roads.

From the 1840s the canals gradually lost ground to the railways, but even so some areas still had a sizeable canal traffic in coal until well into the twentieth century. One of the last collieries at which coal was loaded direct into canal barges was Stafford Colliery, Stoke-on-Trent, which closed in 1969. The loading boom may still be seen at the time of writing.

We shall return to the subject of underground transport in Chapter Four. Perhaps we should make it clear here, however, that

1 (*above*) Bell pits at Stretton, Derbyshire, probably sixteenth century. The author sets the scale. 2 (*left*) Detail of eighteenth century roadway at Stretton showing prop and lid

3 (*above*) Tandem headgear at Brinsley (dismantled 1970). 4 (*left*) Headgear at Upcast Shaft, Brora Colliery, Sutherland, showing cage

underground tramways in the late eighteenth and early nineteenth centuries were generally used only on the main roads of a colliery. The coal was transported for fairly considerable distances in wheelless corves (sometimes mounted on sledges) and in some pits there were no rails at all. Rails were rarely used at all before about 1760, except in the Shropshire drift mines. From 1760 wooden rails began to come into general use at the larger collieries for main road haulage, and M. Jars in 1765 remarked on the horses being taken down the pit for this purpose in the Great Northern coalfield.[14] About 1790 a well-known mining engineer, Curr, introduced light cast iron plates at the Duke of Norfolk's mines near Sheffield. His rails 'were commonly 6 feet long, 3 inches broad in the trod, and ½ inch thick; the margin, or flange, being 2 inches higher than the plate'.[15] The wooden sleepers were about three feet four inches long and the trams held five and a half to six cwt of coal.

As we have indicated, iron rails soon replaced wooden ones except in Shropshire. A Derbyshire witness giving evidence to the Children's Employment Sub-Commissioner in 1841 said how much this had lightened the work of the haulage lads who, in that county, had to push the tubs along the main road. Thomas Wilson expressed the same view in his north-country poem 'The Pitman's Pay':

> God bless the man wi' peace and plenty
> That first invented metal plates,
> Draw out his years to five times twenty,
> Then slide him through the heevenly gates.[16]

A few collieries had an even more efficient means of underground transport than iron rails, namely subterranean canals. These were collieries sunk on hillsides where a water-level could be cut through the hill, thus providing both drainage and transport. Probably the first such canal, and certainly the most famous, was that at Worsley Colliery, owned by the Duke of Bridgewater. This was part of the Bridgewater Canal from Worsley to Manchester, engineered by James Brindley, which had the effect of halving the delivered price of coal in Manchester. There had previously been a sough draining the Worsley workings into the Worsley Brook. The original survey for the canal was the work of John Gilbert, and it was probably his idea to use an enlarged water-level for a dual purpose.

By the end of 1759 some two miles of canal had been completed above ground, plus 150 yards of the underground branch. The underground water level had a main channel and several branches. The coal was dragged in corves from each coal face to the nearest branch, so that the canal took the place of the wagon-ways used at other mines. Later, two more water levels were driven to serve lower seams. One of these was 56 yards and the other 83 yards below the main level. Coal was raised up staple shafts from these lower levels to the main level by winches. Later still a fourth level, higher than the main level, was driven and worked similarly by staple shafts and winches until the completion of an inclined plane between these two levels obviated the need for them.

This network of underground waterways totalled forty-two miles and some parts of the system were in use for transport until 1887. Mr Charles Hadfield travelled part of the main level in 1961 and found that the roof, originally eight feet high, was down to four feet in places because of subsidence.[17] A film in the N.C.B.'s Mining Review Series features this canal. Now that all the mines in the district are closed, the underground section of the waterway is no longer accessible; but the entrance is not sealed because of the possibility that gases might accumulate and seep through old shafts to the surface.

Boats were propelled through the Worsley levels by 'legging' on the roof. This was, of course, the usual way of working through low tunnels on canals. The same means of propulsion was adopted, for example, at Butterley, Derbyshire. Several of the Worsley boats (called 'starvationer' boats) have survived, and one will shortly be exhibited at the Lound Hall Mining Museum.

There were similar subterranean canals at a number of mines in Wales and Shropshire (the best known being the Navigation Level near Donnington), and one or two in Derbyshire. One shown on old mine plans was driven from the Chesterfield Canal at Staveley for about a mile and a half to tap coal and ironstone workings in the Westwood area. In places it was seventy feet below the surface. This underground waterway was six feet high and six feet wide with a depth of water of two feet. The barges used were twenty-one feet long. The Worsley boats, by contrast, were fifty feet long.

Some mention has been made of staithes. These were coal wharves at harbours on the banks of the Tyne and Wear and at coastal ports on the Northumberland–Durham coalfield. The coal was, as we have seen, transported from the pits along wagon ways. In the early days coal was stacked at the wharf in large heaps covered with timber roofs to prevent it from becoming too wet. It was transported in wheelbarrows along narrow earth causeways jutting out a little way into the water, and then tipped into small boats called keels. The keels carried the coal out to the coastwise vessels, often called colliers, which conveyed it to London and other places.

In the early eighteenth century, piers were built out into the river to facilitate the tipping of the coal into the keels. On the end of the piers wooden chutes (called 'spouts') were erected. These speeded up the loading of the keels but they also caused considerable degradation of the coal. A device to reduce degradation was invented by William Chapman of Newcastle about 1800 and first employed, at Benwell Staithes, in 1808. This was the coal 'drop'. A contemporary observer explained that the drop was:

> a square frame hung upon pulleys, and counterbalanced by back weights. The loaded waggon, together with the square frame, descends by its own gravity to the hatchway of the vessel, delivers its coals, and, in turn, the empty waggon is returned by means of the balance weights, the motion being in both cases regulated by a brake wheel. A man is lowered down with the waggon, whose business it is to unhasp its moveable bottom, and thereby let the coals drop into the hold of the vessel.[18]

Some of the staithes were, by this time, being built well out into deep water so that coal could be loaded into seagoing vessels without the intervention of keels. The first such staithes on the River Wear were built in 1812 at Sunderland. There were three spouts at first, but spouts were replaced by drops in 1820. When these staithes were opened, the keelmen feared that they would be put out of work; and in 1815 part of the structure was pulled down during a riot. A remarkable picture of staithes at Wallsend showing drops and a spout is included in Hair's *Sketches*.

Some modern staithes still have spouts, for example, Amble Staithes, Northumberland, a photograph of which is shown in Atkinson's *The Great Northern Coalfield*. The last coal drop to remain *in situ* was at Seaham Harbour, but this has now been removed and is stored preparatory to being re-erected at the Northern Regional Open Air Museum.

As we have seen, in the early days it was usual to have considerable quantities of coal stacked at the staithes awaiting shipment. This ensured that shipmasters would not be kept waiting for coal. Similarly, inland collieries maintained stacks of coal at their wharves. This was particularly important before the canal era when wharves were maintained on the outskirts of towns to be drawn on in winter when the state of the roads prevented the transport of coal. Even during the canal age, coal was still stacked at the wharves so as to make sure that a sale was not lost to some other colliery owner through want of ready supplies. And small coal sometimes remained on the canal bank, to become consolidated and, in time, covered with soil. Some of these heaps of slack have been discovered and exploited in fairly recent times when coal has been in short supply.

Comparatively little coal is nowadays carried on inland waterways. Similarly, steam locomotives belong to the past. As with main line traffic, colliery traffic is now hauled by diesel locomotives in many areas; even so the N.C.B. probably has more steam locos in use than all other users combined. Some of the colliery tank locos in use thirty years ago had an old-fashioned look about them, with their long chimneys. Few of these 'coffee-pot' engines survive.

Perhaps the most interesting colliery locomotive held by the N.C.B. is preserved at the Philadelphia Workshops, Durham. This, the *Bradyll* loco, was built by Timothy Hackworth in 1837 and was one of four used at South Hetton Colliery. It was in use as a locomotive engine until 1875 and was then used for many years as a snow plough. Timothy Hackworth had worked as a young man on the famous *Puffing Billy* which came into use, as we have seen, on the Wylam Colliery Railway in 1813 and which is now preserved at South Kensington. On a locomotive which he built in 1827, Hackworth used two cylinders to drive the same axle-tree and also forced the waste steam through a narrow orifice into the chimney thereby increasing the force of the steam blast. These improvements were subsequently adopted by other engineers, including George Stephenson.[19] Other old colliery engines displayed in museums include the Killingworth Colliery locomotive built by Robert Stephenson & Co. in 1830 and in use until 1881, which is at the Newcastle Museum of Science and Engineering; and *Wylam Dilly*, built in 1813, which is at the Royal Scottish Museum, Edinburgh.

The more powerful locomotives built from this time were able to haul heavy loads up gradients on smooth rails without the aid of a rack-rail.

So closely connected were the early railways with the mining industry that it has been necessary to outline some of the important railway developments here, but we must now return to our proper study.

Shafts and Winding

Early methods of mining, by drifts and bell pits, have already been described. These early coal mines were all on exposed coalfields where the seams were seen to outcrop on to the surface. There was therefore nothing speculative about them: the coal was known to be there.

With mines away from the outcrop the presence of coal might be suspected but was not certain. Sinking was therefore a speculative business until Huntington Beaumont demonstrated his 'art to boore with iron rodds to try the deepnesse and thicknesse of the cole'.[1] Exploratory boring, which Beaumont took with him to the North-East coalfield in 1605, was especially important in that district. Because of the pull of the London market, the collieries of Northumberland and Durham were comparatively large-scale enterprises, their shafts deeper (since the shallow seams were soon worked out) and their workings more extensive than collieries in other parts of the country. In 1708 boring was said to cost fifteen to twenty shillings a fathom for depths of thirty to sixty fathoms, whilst sinking a shaft cost at least fifty or sixty shillings a fathom.[2] Besides indicating the depth and thickness of the coal seams, boring also exposed the nature of the superincumbent strata, whether it was waterlogged, whether there was running sand or particularly hard rock, and so on. This information enabled the sinkers to make appropriate preparations.

The method of boring described by the author of *The Compleat Collier* in 1708 remained virtually unchanged for a century:

> We have two Labourers at a time, at the handle of the bore Rod, and they chop, or pounce with their Hands up and down to cut the Stone or Mineral, going round, which of course grinds either of them small, so that finding your Rod to have cut down four or six Inches, they lift up the Rod, either all at once, as there is

> conveniency for its Lift; or by Joynts, fixing the Key, which is to keep the Rod from droping down into the hole... and taking off the cutting Chissel, puts, or screws on the Wimble or Scoop which takes up the cut Stuff be it what happens; and so by sight of the Stuff... you plainly discover of what Kind it is... and so consequently what follows to any depth; for by Addition of Joynts, (which we screw into the Rod,) we can descend to any depth....

A new system, which allowed cores to be withdrawn from the boring, was invented by James Ryan in 1804. Steam power, first applied to boring by Richard Trevithick, who invented a rotating boring machine, both speeded up and cheapened the exploratory process. Boring by hand nevertheless continued at some Lancashire collieries into the twentieth century as D. Anderson has shown in a recent article on Blundell's Collieries: 'More modern methods of boring were invented, especially for boring deep holes, but in the 190 years existence of Blundell's collieries (1776 to 1966) only the ancient hand percussive method was used for boring vertical holes from the surface.'[3]

Shaft sinking could be either simple or hazardous depending on the depth of the sinking and the nature of the strata. The shallow pits of the Midlands, many of which were no more than fifty or sixty feet deep as late as the early nineteenth century, presented few problems except where water was encountered. They were usually circular in shape and anything between five and twelve feet in diameter, six to eight feet being normal. Sinking was originally carried forward with simple hand tools: picks, shovels and wooden wedges. Later, boring rods were also used. One of the earliest references to the use of explosives in sinking concerns a rectangular shaft sunk in the Halifax district to a depth of 210 feet in 1749.[4]

The last of the Blundells pits to be sunk using hand drilling methods was Prince Pit, Pemberton Colliery where sinking commenced in 1898. Anderson tells us that: 'The sinkers' tools used were almost exactly those in use at the beginning of the Nineteenth Century.' They included heavy picks, shovels, wedges, and 'sledges' (heavy hammers). Steel drills with cutting and striking ends were used:

> One man held and turned the drill after each blow and the other two struck it with sledges one after the other. Jumpers were used

with two hands, the operator raising it and letting it fall with considerable force in the hole he was drilling. This was used for soft and moderately hard rock. A scraper was used for removing chippings from the hole.

The loosened rock was filled into a large bucket called a 'hoppet' or 'kibble' in which it was raised to the surface. The hoppet swung freely in the shaft, there being no temporary guide rods or ropes.

Square or rectangular shafts were common in some coalfields. They were preferred because they were easier to line with wooden boards than circular or oval shafts. Rectangular shafts were sunk in Scotland well into the nineteenth century. The Forest of Dean was another area where small rectangular shafts were common.

An eighteenth-century shaft, approximately five feet three inches square has been exposed in section on the Swan Farm (Shropshire) opencast site where it could still be seen in September 1969. This shaft, 100 feet or so deep, is roughly lined with timber boards.

In Northumberland and Durham the usual practice in the eighteenth century (and probably earlier) was to sink the shaft four-square through the loose earth, altering for a short distance to an octagonal shape to facilitate the transition to a circular shape when the hard stone was encountered. Water was held back by the use of 'tubbing', that is wooden staves formed in the same way as a tub. Cast iron tubbing was later introduced, being used in the north from about 1795. The first cast iron tubbing used in the Midlands was at Snibston Colliery, Leicestershire, sunk by George Stephenson in 1841–3. This can still be seen in the upcast shafts.

The Hartley Colliery shaft, sunk in 1845 and the scene of a disaster in 1862 which we shall describe later, was apparently of the traditional Northern shape. The shaft was said to be twelve feet three inches in diameter, but it is undoubtedly square at the top. It is shown to be so in a contemporary engraving, and the square stone wall which now surmounts it appears to be an extension of the original shaft wall.[5]

The Hartley shaft was 200 yards deep and was only walled for the top few yards. Below that, as an Inspector reported: 'The strata where considered liable to give way, were secured by a lining of planks kept in their places by round timber cribs or curbs of from

4 to 5 inches scantling.'[6] It would have been much safer to line the whole shaft with bricks or stone, but this would have added to the already high cost (£3,600) of sinking. Hartley had only the one shaft, in common with many of the large collieries in Northumberland and Durham. This shaft was divided vertically by a brattice of timber planks. The southern half of the shaft was the downcast (i.e. it had a downward flow of air) and it held the cages used for winding. The northern half of the shaft was the upcast, the upward flow of air being induced by an underground furnace, and it also accommodated the pumping gear (which would retard ventilation).

Making one shaft do the work of two saved expense but it was, as we shall see, dangerous. Fortunately, single shaft pits were uncommon elsewhere by the nineteenth century, although one prominent north-country mining engineer had tried to introduce them into the Black Country in 1798. He condemned the 'provincial' method of employing two six-foot shafts and proposed instead one twelve-foot shaft divided down the middle with a wooden brattice.[7]

Hartley Colliery was not alone in having unsatisfactory shaft linings. In 1841–2 the Children's Employment Commissioners noted many instances of shafts which were either unlined or lined inadequately. This was true of many of the rectangular shafts common in backward districts. A Mines Inspector expressed the view in 1851 that 'the sinking of square or oblong pits instead of circular ones is to be deprecated, as the former are more insecure and more costly than the latter'.

In support of his view he commented on a case at Bedminster, near Bristol, where forty miners were buried alive for thirty hours when the side of a square shaft collapsed.

In the Nottinghamshire–Derbyshire coalfield, many pits were lined with dry (i.e. unmortared) bricks supported on a wooden curb. It was cheaper to lay bricks dry, and they might then more easily be used again when the shaft ceased production.

Let us consider now how the coal was sent up the shaft. With a very shallow bell pit, it was possible to throw the coal out of the pit with a shovel. Below that depth, the coal was usually hauled out manually in a bucket or basket on the end of a rope, but in Scotland and some other places coal was carried up the shaft by bearers. This may have been done in south Staffordshire at an early date because wooden platforms were found at intervals in a shaft exposed during construction of the M.6 Motorway near Walsall, indicating the use of ladders. We shall return to this subject in Chapter Four.

The hand windlass was used from a very early date, both for winding coal and also for winding water out of the shaft (Fig. 1). In parts of the east Midlands it was usually known as a wallower or 'waller'. Some wallowers had a handle at each end so that two people could operate them. This would no doubt be necessary at deep pits. Windlasses were in some cases used to wind people into the pit. Indeed the Children's Employment Commission Report of 1842 has a contemporary illustration showing two children, sitting astride a short wooden bar (or 'horse') fastened to the rope being wound down by a windlass operated by an oldish-looking woman. However, most miners considered ladders to be a safer means of travelling the shaft and used them if they were available.

The barrel of a very old wallower was found in 1969 at the bottom of a shallow shaft on the Stainsby Hag opencast site near Heath, Derbyshire. This is now held at the Lound Hall Mining Training Centre. Quite a number of nineteenth-century windlasses of various patterns survive, and there are a few still in use in the Forest of Dean.

The next development was the horse-driven cog-and-rung gin introduced in the first half of the seventeenth century (Fig. 5). This device had a rope drum, very much like the cylindrical part of a windlass, suspended directly over the shaft with a hazel basket or corfe on either end of the rope. The horse track was around the mouth of the pit, and as the horse walked round he drove a horizontal cogged wooden wheel which was mounted on a vertical axis. The cogs on this wheel meshed on cogs mounted on the rope drum. Winding in the reverse direction was effected by turning the horse round.

A very interesting watercolour of a pit top showing a cog-and-rung gin being operated by a single horse was painted by John Bewick in about 1775. A photographic reproduction of this is in Frank Atkinson's excellent short study, *The Great Northern Coalfield.*

An improved horse winding device, the whim-gin, was invented towards the end of the seventeenth century and it soon replaced the cog-and-rung gin. This new winder, the whim-gin (or whimsey) had a horizontal rope drum mounted on a vertical shaft (Fig. 6). The drum was at a distance from the pit mouth, the rope being suspended

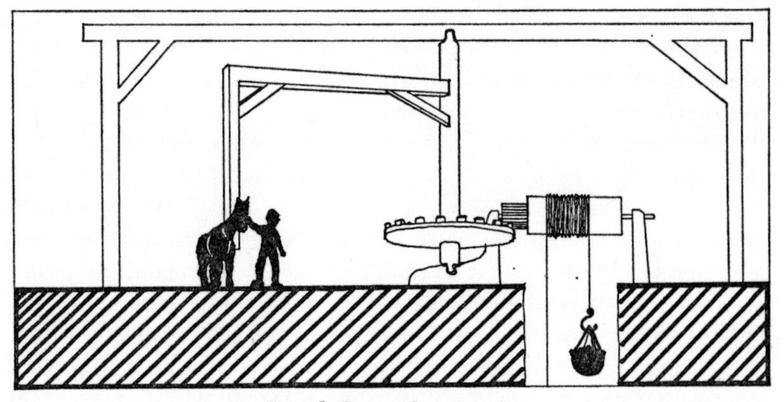

Fig. 5 Cog-and-rung gin

over pulleys. The drum of a whim-gin could be made larger than that of a cog-and-rung gin, thus giving more power. Some gins were worked by one horse, others by two or more horses depending on the depth of the shaft and the size of the load to be raised. And of course the horses were well away from the pit mouth, since the horse track no longer needed to encompass the shaft. Whim-gins were used at some small mines in the latter part of the nineteenth century, and quite a few survived particularly in Shropshire into the twentieth, leaving, in some cases, the circular walk of the horses as evidence.

Typically, the rope drum was about nine or ten feet in diameter

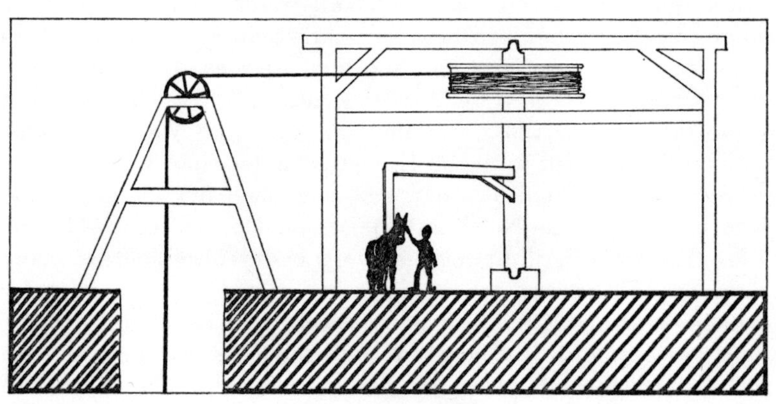

Fig. 6 Whim-gin

and was mounted on a substantial shaft some twelve or fifteen feet long which had an iron spindle at either end. One spindle fitted into a hole in the floor whilst the other fitted into a hole in the cross beam. The cross beam, usually about forty feet long, was rather like the cross-bar of the goal posts used in association football, and was similarly supported by uprights. In the simplest gins, the horse arm, ten or twelve yards in length, was morticed through the shaft of the rope drum just below the drum. Specially designed harness was used to attach the horse to the end of the arm. The pit head frame, some distance away from the gin frame and similar in design, carried two pulley wheels two or three feet in diameter over which the rope passed down the shaft. Across the uprights of the pit head frame was a horizontal striking bar to support the banksman whilst drawing the loaded corve to the shaft side with a long hook.

An improved whim-gin which employed gearing and was driven by eight horses raised six cwt of coal from a depth of 200 yards in two minutes at Walker Colliery, about 1750. In coalfields other than the north-east, however, one- or two-horse gins sufficed.

In Shropshire another type of horse winding device was sometimes used. A pulley was erected over the shaft and the horse walked away from this down a slope drawing the winding rope and thus raising the box or basket of coal. This was known as the jack-weight system, and the pits using it were called horse-gear pits.

Water wheels had been used in Germany for drawing the ore out of metal mines from the middle of the sixteenth century, and Huntingdon Beaumont had used both water- and windmills for mines drainage early in the seventeenth century. However, there appears to have been no attempt made to use water power for winding coal until the early eighteenth century, when a water-whim came into use at Alloa, Scotland; but even this was an isolated example. It was not until the 1770s that waterwheel winders were introduced into English mines, and even then few of them came into use outside the Northumberland and Durham coalfield. One exception was at Griff Colliery in Warwickshire, the home of many technical innovations, where Smeaton erected a waterwheel for winding coal in 1774. These waterwheels were mounted vertically on a horizontal shaft (Fig. 7). Around the edge of the wheel were double buckets, some facing one way and some the other. Above the wheel were two channels, one on each side of the wheel, through which the water ran to operate the wheel. The channels were so positioned that when one was open the water ran into the buckets facing in one direction; whilst when the other was open the water ran into the buckets facing the other direction. Thus, the motion of the wheel was reversed by cutting off the water from one channel whilst releasing it from the other. A rope drum, poised directly over the shaft, was mounted on the same horizontal shaft as the water-wheel so that the corves were alternately raised and lowered as the direction of the water flow alternated.

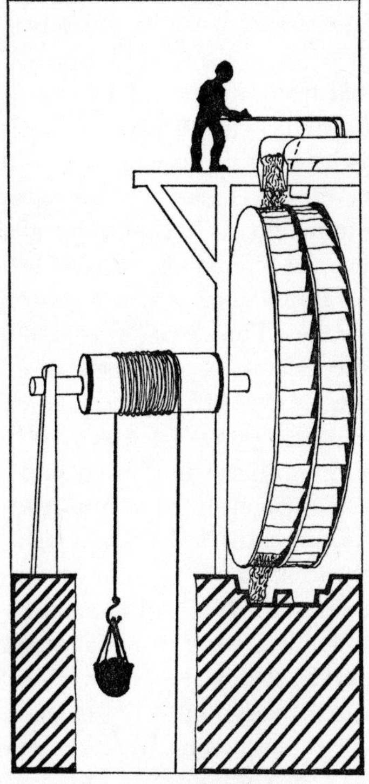

Fig. 7 Water whim

At several collieries in the north-east, steam engines were used to raise water for the water-gins and Smeaton expressed the view that this was the best way to harness steam for winding coal because of the difficulties inherent in adopting steam engines to rotary motion. In fact, the first steam winder, a Watt engine, was erected at

Walker Colliery on Tyneside in 1784, and within a few years engines of both the Newcomen and Watt types were in use for winding,[8] so Smeaton's view was disproved.

In South Wales some mines employed a device known as the water-balance winder. Here, coal was drawn up the shaft in one cage by the weight of a tub, filled with water at the pit mouth, in the other. When the water-tub reached the pit bottom, the water was drawn off through a valve in the tub's base. A tram of coal was then put on top of the empty water-tub; and drawn up the shaft in the same way as before. An example of this device cited by Morris and Williams in their interesting book on the South Wales coalfield was at Cwm Bargoed pit where, in a twelve-hour shift, it was possible to wind 300 tons of coal from a depth of 154 yards. Another example, from the Forest of Dean, is on display at the National Museum of Wales, Cardiff. It is interesting to note that water balance winding systems are now being developed for use in emergencies (e.g. an electric power line failure) to wind men out of the pit.

The water-balance mechanism could also be used for winding at drift mines; and the remains of one such water-balance tower may still be seen at Blaenavon. This device was popular for shallow mines on account of its low capital cost and ease of operation, but by the third quarter of the nineteenth century the deeper seams being worked called for a more powerful means of winding; and steam engines came into general use in the coalfield. As we have seen, steam winders were in use elsewhere almost a century before. It is unfortunate that the early steam-driven winders were usually called whimseys (variously spelt) so that it is sometimes difficult to tell whether a reference is to a whim-gin or steam engine.

The early steam winders were crude affairs and many owners of small collieries preferred still to use whim-gins. Engines often broke down, and it sometimes took months to get them repaired because of the shortage of skilled engineers. Wilkes, who installed the first steam winder in the east Midlands in 1787, had exactly this experience. Attempts to produce rotary motion from a Newcomen engine were made from about 1768, but without any real success until the adaptation of the crank about 1780. Where a steam winder was installed it was usual to retain a whim-gin as a standby. Standby whim-gins in Durham were popularly called 'crabs' in the nineteenth century, presumably because of their shape.

Whim-gins were also preferred for sinking during the first half of the nineteenth century at the fairly shallow pits of the east Midlands and elsewhere and were still preferred by some later in the century. One of the last whim-gins to be in regular use was built as a sinking-gin. The pits of the firm concerned (Coke and Company) were all worked with steam winders when the Children's Employment Sub-Commissioner visited the Derbyshire district in 1841, and had been so for some years. However, a whim-gin was used to sink the new Langton Colliery, opened in 1844. This gin was then moved to what had been the Number 1 pit of the Old Pinxton Colliery but which became a pumping shaft after Langton commenced production. The gin was in regular use for shaft inspection work until about thirty years ago. It remained *in situ* (though in a ruinous state) until the building of the M1 Motorway when it was removed to the N.C.B.'s South Normanton Training Centre, where it was completely rebuilt. At the present day it lies in bits and pieces at the City of Nottingham's proposed Industrial Museum at Wollaton Hall. A similar one-horse gin, formerly used at Rothwell Colliery, Leeds, is featured in Kenneth Hudson's pioneer book on Industrial Archaeology. I. J. Brown recalls seeing a new whim-gin at the Court Pits, Shropshire, used in the surveying of old shafts about 1944. By contrast with the whim-gin used to sink Langton in 1844, a shaft at Florence Colliery in Staffordshire was sunk with a powerful steam-powered sinking engine which is still used for shaft work and as a stand-by winder in case of emergencies. This was made by Tickle Bros of Wigan in 1902, and is now located at Stafford Colliery, Stoke-on-Trent.

The early steam engines used for winding were single-cylinder vertical engines. They had grab hand reversing gear and were usually between seven and fifteen horse power; although by 1850 some large collieries in the north-east were employing engines of up to 50 h.p.

A miner's description of a winder in use in the early nineteenth century, although non-technical, is evocative:

> The beam of wood; fly-wheel and drum, with horns, form the chief ornament; no steam indicator or brake are to be seen; but there is a poking handle, like an iron walking-stick, which on every move makes a rattle and clink-clank enough to frighten a

statue. Yet often as it brought its load to the surface, the banksman would have to run and give the fly-wheel a 'lift' with his shoulder.[9]

Fig. 8 Atmospheric engine, South Staffordshire
(*sketched by I. K. Griffin after Midland Mining Commission Report 1843*)

Contrary to what has been supposed by some historians, many of the early steam winders were atmospheric engines with open-top cylinders based on Newcomen's design. These were much smaller than pumping engines. By 1790 Coalbrookdale ironworks were doing a regular trade in these small engines with cranks for winding. They used a great deal more fuel than Watt engines but since they used small coal which was otherwise almost unsaleable this was not an important consideration. After Watt's patent of the separate condenser, Newcomen engines were also sometimes produced with a separate small vessel attached to the cylinder in which the steam was condensed thus improving efficiency. These were known as 'pickle pot' condensers. The large collieries of the north-east recognized the superiority of the Watt double-acting engine, but Newcomen engines continued in use at small pits in other parts of the country well into the nineteenth century. The Farme Colliery, Rutherglen, was among the last to employ Newcomen winding engines. One engine which was commissioned there in 1810 was in use until 1915.[10] It is now in store at the Glasgow Art Gallery and Museum. This engine has a forty-two inch cylinder with a

5 (*above*) Atmospheric pumping engine, Elsecar, Yorkshire. Built in 1795. 6 (*below*) Flat rope vertical winder, Wearmouth

7 Whim-gin from Pinxton, built *c.* 1840

8 Old Glyn Engine, East Wales

stroke of five feet eight inches. One rope drew from a depth of sixty yards and the other from eighty-eight yards. Originally, the engine was also used for pumping. There are two photographs of this engine in the *Proceedings* of the Institution of Mechanical Engineers for 1903 (Fig. 9). The maker of the engine was John McIntyre and the castings were supplied by the nearby Camlachie Foundry. A second atmospheric engine, for winding, was added in 1820; and another for pumping in 1821. These two engines were dismantled in 1888.

Fig. 9 Newcomen Winding engine erected 1810 and working until 1915—Farme Colliery, Rutherglen (*sketched by I. K. Griffin after I.M.E. Proceedings, 1903*)

Another late survival of the atmospheric engine was photographed at a Shropshire mine in 1917.

One reason why the early Newcomen winder was preferred by many to the Watt engine is that it was simpler and more reliable. The Watt engine was superior in design and in principle; but with the low standard of workmanship and unsatisfactory materials available, the simpler atmospheric engine worked better. The first steam winder used in the east Midlands installed by Joseph Wilkes at Oakthorpe in 1787 to which we have referred was, indeed, a Boulton and Watt engine; and Wilkes's experience with this must have been a salutary warning to other local colliery owners. From the beginning trouble was experienced, and after about six months the winder broke down altogether, allegedly because of bad workmanship, although Boulton and Watt's engineer would not accept the accusation. After a repair the engine worked spasmodically; for example, it was out of action from late 1796 until February 1798 or later.

Farey counted fifty or more steam 'whimseys' in Nottinghamshire and Derbyshire in 1811. Most of these were atmospheric engines. Haystack boilers were used for these early engines; but later the wagon-boiler superseded the earlier type.

An interesting winding engine erected at Wearmouth in 1868 was in use until 1960 when it was dismantled. This is a single-cylinder vertical engine with a chain counterbalance and flat ropes. It is now in store awaiting re-erection at the Northern Regional Open Air Museum After 1868 the great majority of winding engines were horizontal, although there were two vertical engines made by Wren and Hopkinson installed at Stafford Colliery, Stoke-on-Trent in 1875, which are still in use at the date of writing. Coal production has now finished at Stafford, however, and no doubt these unique engines will soon be scrapped. They have been offered to museums, but because of their bulk it would be difficult to rehouse them, quite apart from the expense involved in dismantling, transporting and re-erecting them.

These two winding engines and the tall engine houses which contain them are identical. Coal was drawn from both shafts: Homer (upcast) and Sutherland (downcast), lately from a depth of 900 yards, although several shallower seams have been worked. The engineman sits at decking level to operate the controls and the rope drum is eighteen feet above him. The engines were fitted with the 'Mellins' piston valve in the early 1920s. The cylinders are thirty-six inches in diameter with a 72-inch stroke and they use steam at ninety pounds per square inch.[11]

Another vertical engine still intact is mentioned in Atkinson's

The Great Northern Coalfield. This was in use at Beamish Colliery until 1962. It was built by J. and G. Joicey of Newcastle-upon-Tyne in 1855. The headstocks at Beamish, illustrated in Atkinson's book, resemble those at Wearmouth. Instead of a vertical structure standing four-square free of the engine house, as with the typical headstocks, this one has two wooden legs standing over the shaft and two arms connecting the top of the headstocks, cantilever fashion, to the front of the engine house. The arms look almost horizontal in the picture. At Stafford the usual four-square headstocks is strengthened by girders apparently attached to the engine house but low down on the wall, presenting a far more normal appearance.

Another vertical single cylinder winding engine still intact is at Old Glyn mine, near Pontypool, where there is also a beam engine, formerly used for pumping, in a roofless engine house erected in 1845, and built from local stone. A photograph of this, taken by D. E. Bick of Cheltenham, appeared in *Industrial Archaeology* in November 1969 and is reproduced here. These two engines were made at Neath Abbey Works. The winder employed flat ropes; and a flat rope reel is still *in situ.* The Old Glyn engines, and the engine house, are scheduled as Industrial Monuments (see Appendices B and C).

Yet another old vertical winding engine is at Cadley Hill Colliery in south Derbyshire and another is at Ireland Colliery in the north of the county. There are still quite a few steam winding engines with horizontal cylinders in all coalfields, but some go out of use every year and they will be rare in five years' time. One such engine, more interesting than most, was the subject of a film entitled *Winding at Old Mills*, made by the N.C.B. Film Unit in 1965. Since then this engine has gone out of use and the colliery at which it was situated has been closed. The film shows the engine actually at work, and is therefore a much more satisfactory record than either a written description or still photographs. Old Mills was one of the few collieries in the Somerset coalfield. The winding engine was built in 1861 by W. Evans of Paulton Foundry, Somerset, a firm which has long since gone out of business, and it was known, affectionately, as the 'Old Lady'.

The 'Old Lady' has two horizontal cylinders of five foot stroke and twenty-six-inch bore; and the steam pressure was forty-five pounds per square inch. The rope drum is twelve feet in diameter. The coal was drawn from a depth of 375 yards on a winding rope of $1\frac{1}{8}$ inch diameter. This engine had a number of interesting features. The band type brake was foot operated and it was powered by its own small steam engine. With this type of brake, depressing the brake pedal cuts off the steam and this activates the brake immediately. The brake therefore comes on with a very sharp jerk. An unusual feature of the Old Mills engine was the exhaust arrangement. With most steam winders, there is only one exhaust pipe to be seen, but here there were two sticking through the stone wall of the engine house, emitting puffs of steam alternately. This engine is to be preserved at the Bristol Museum.

An example of a horizontal winder which has recently become disused and is to be broken up for scrap but is still intact at the time of writing is at Handsworth in South Yorkshire. It has twin cylinders, twenty-six inch bore, sixty inch stroke; with a sixteen-foot diameter parallel drum, wood lagged. This engine is fitted with a Walker overspeed device. When working it drew coal from the Parkgate Seam (400 yards deep) at sixty winds per hour.

It is, to put it mildly, a pity that all these old steam winders are either being scrapped or, at best, sent to museums. Steam is essentially alive. A steam engine in a museum is dead. Ideally, an engine should be preserved *in situ* and in working condition. Unless this is done soon it will be too late.

While most winding engines were specially built for the purpose, some were adapted from other uses and a few of these makeshift engines are still in use. For example, there is a double-cylinder horizontal engine at Bilsthorpe Colliery, Nottinghamshire, which was built in 1886 by Thornewill and Wareham of Burton-on-Trent as a marine engine. The Stanton Ironworks Company used it as a standby at Pleasley Colliery before transferring it to Bilsthorpe in 1925. It was used there as the sinking engine and has since been used at the No. 2 Pit mainly for drawing men and materials. It uses steam at eighty pounds per square inch.

The earliest electric winders came into use about 1906 but it is only since 1947 that they have been widely adopted in place of steam winders. This trend can be expected to continue. Some electric winding engines are able to work automatically, though it is still necessary to have an engineman on duty.

Until about 1840 all winding ropes were made of hemp, though chains were sometimes used instead of ropes. The chief drawback to winding chains was that they could break without warning, whereas a hemp rope showed signs of wear first. Chains were popular in Shropshire, however, where there was a substantial local chain-making industry. Triple-link hand-forged winding chains continued in use at some small mines in Shropshire until about 1935, though winding chains were little used elsewhere after 1840. The winding chains used in the twentieth century were endless chains operating in much the same way as a bicycle chain. A photograph showing the pit top arrangements of one of these 'chain pits' (Bridge Pit, near Ashton-under-Lyne) is to be found in D. Anderson's interesting study of Blundell's Collieries.

The usual method of winding coal out of the pit before 1840 was for the pit bottom man (the hanger-on) to attach the corve to a hook on the rope or chain. At the same time, the pit top man would be removing a full corve from the other end of the rope and re-attaching it when empty. As the full corve ascended the shaft, therefore, the empty one descended. At some small pits there was only a single rope in the shaft but this was unusual in the early nineteenth century. At large collieries, instead of one corve there were several, each hung on a separate hook. A sketch of Hebburn Colliery 'C' pit by T. H. Hair, originally published in 1839, shows three corves hanging from the rope.

A Durham miner, who started work in 1837 at nine years of age, described the system and explains what happened when men were to ride the shaft:

> Instead of a cage a long chain was attached to the rope on which at equal distances were placed three hooks. From each hook was suspended a large basket made of strong hazel rods, closely interlaced and twisted round, as they were firmly fastened to an iron bow. Such baskets were called 'corves', and could carry from twenty to thirty pecks of coal. Generally, when men or lads were to ascend or descend, the corves were taken off and the hook was passed through a link of the chain, thus forming a large loop, in which two men each placed a leg. They grasped the chain with their arms, and a little boy was then set astride their knees. He grasped the chain with both hands, and they held him to themselves with their free arms. Then they were lowered a little till the loops

> above were similarly occupied, and sometimes the space between filled with lads clinging with arms and legs to the chain. Above the top loop ten to twenty or more lads would catch the chain, till fathoms of rope and chain, covered with human beings, dangled over the dark abyss.[12]

This was a large colliery, of course. At the small collieries of the Midlands it was more usual for eight, ten or fifteen people to ride the rope on one draw. Whether there was a 'bantle' of eight or forty people, however, the ride would be singularly uncomfortable because there was nothing to hold the rope steady.

Indeed, it was by no means uncommon for someone to fall off the rope. Occasionally, someone would do so as a result of being affected by 'black-damp'. Again, where the shaft was lined with loose bricks a section of lining sometimes gave way and could throw a miner off the rope. Loose bricks or debris from the surface could hit a rider on the head. To avoid this particular danger, at some collieries an iron 'bonnet', shaped like a dustbin lid, was fitted on to the rope above the men during man-riding operations; but this brought a new danger where, as sometimes happened, coal was wound up while men were being wound down, because if the ascending corve fouled the bonnet the men might be caught unawares and tipped off the rope.

In the 1840s, cages were widely introduced, and by 1860 most collieries of any size had them. The early cages had open tops, but it soon became usual to make them with 'bonnets'. A system of guide bars to hold wooden trams steady in the shaft had been invented by John Curr of Sheffield in 1787 but had not been widely adopted. About 1835 T. Y. Hall invented a greatly improved system. Here, wooden guide rails ran down the side of the shaft, and iron cages were fitted with shoes which ran on the rails in much the same way as a wagon runs on railway track. Instead of hazel corves, the coal was now loaded into small wheeled trams which carried it to the pit bank. This new system was recognized as a great improvement in efficiency as its rapid adoption testifies.[13] And for man-winding the change—from riding an oscillating rope to riding in a cage held steady by guide rails—was revolutionary. Even so the old system lingered on in some of the small, backward collieries of south Staffordshire, Worcestershire and Shropshire into and, in a few cases, beyond the 1870s. Similarly, corves were used at the William pit, Whitehaven, until 1875. One of the surviving Whitehaven corves has been reserved for the Lound Hall Mining Museum.

Another great improvement adopted widely in the 1840s was the iron wire winding rope. However, the conservative-minded colliers of the north-east at first refused to ride on wire ropes because they considered them to be unsafe. After mid-century steel replaced iron. In Shropshire there was in this respect as in so many others a late survival, hemp ropes being used at some small mines into the twentieth century. Lilleshall No. 1 (Grange) Colliery, had a flat hand-stitched winding rope until about 1950.[14] This was a steel wire rope, but was constructed in much the same way as the flat hemp ropes of the mid-nineteenth century.

One of the great dangers of riding the shaft was of being 'overwound'. This sometimes occurred when the winding engineman failed to brake early enough. The cage might then be carried over the pulley wheel. To prevent this several people invented detaching safety hooks. Of these the most widely adopted was that invented by John King of Pinxton in 1866; the similar invention of Ormerod, patented in 1867, ran it a close second. Where a cage was accidentally overwound, the device snatched the cage from the rope and held it firm in the headgear allowing the loose rope to fly into the engine house. A cage which is overwound is still said to be 'kinged'.

The colliery where King's device was tested, Sleights Colliery, Pinxton, had tandem headgear. This kind of headgear was very popular for a time in the mid-nineteenth century, particularly in the Midlands. One of the first collieries to employ this system of winding was Babbington Colliery, Cinderhill, Nottingham, sunk in 1841–3 and still standing, although the original shafts have not been used for coal turning for about twenty-five years. Two seven-foot diameter shafts were sunk close together, and were served by one headstocks shaped like a gantry and having two pulley wheels, one over each shaft. One winding engine worked both shafts with one rope. Each shaft accommodated one cage which latterly had three decks, though the original cages had single decks. As one cage ascended one shaft, the other descended the other shaft. One shaft was a downcast and the other an upcast and they were always called 'Windy Shaft' and 'Smokey Shaft' respectively, although furnace ventilation was superseded well back in the nineteenth century.

Another tandem headgear still to be seen is to be dismantled in the near future. This is at Brinsley Colliery, near Eastwood, where D. H. Lawrence's father worked. It was made famous in the film of *Sons and Lovers*. Tandem headgear was installed in 1872 when the shafts were deepened and widened. For some years Brinsley has been used only for pumping and ventilation, and as an alternative means of egress for a neighbouring colliery. The fan engine was removed about 1965 and both shafts were then converted from upcasts to downcasts. The Brinsley headgear is to be re-erected at the

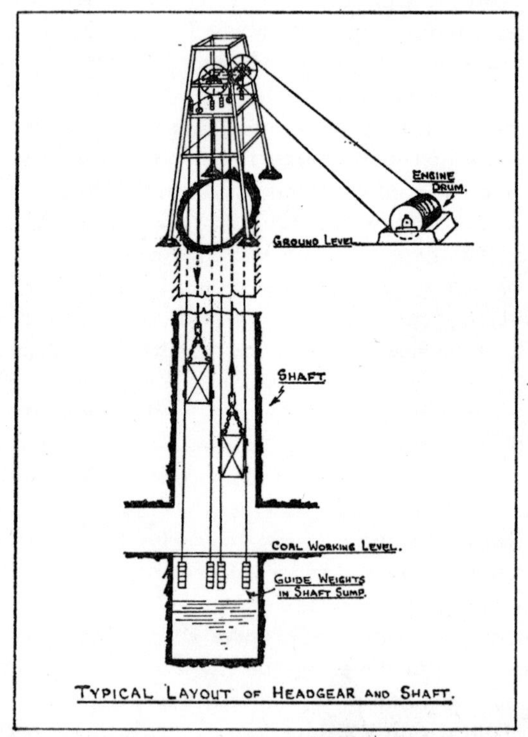

Fig. 10 Typical layout of modern headgear and shaft

Lound Hall Mining Training Centre. One colliery, still in production, with tandem headgear, is Granville (the Grange pit) in Shropshire. This type of headgear was adopted more widely in Shropshire than elsewhere.

Most of the original wooden headstocks on nineteenth-century

pits still in production have been replaced by steel ones, and where pits have closed, their headstocks have usually disappeared. Many of them had well-preserved timbers of pitch-pine which have doubtless been put to alternative use. One colliery where the change from wooden to steel headstocks is about to be made at the time of writing is Wath Main in south Yorkshire.

Small wooden headstocks are occasionally found at shafts not used for coal winding. An example, again taken from south Yorkshire, is to be found at the Elsecar Pumping Shaft. Indeed, the stone engine house, Newcomen pump (dating from 1795) and wooden headstocks make a very interesting group.

In Shropshire an enterprising local society has obtained possession of several old headstocks which they propose to re-erect at Blists Hill, the site of an old iron and brick works associated with a colliery, now used as a branch of the Ironbridge Gorge Museum. The Northern Regional Open Air Museum proposes to preserve similar relics including the headstocks and winding gear of Beamish Colliery to which reference has been made in this chapter.

At South Normanton, on the Nottinghamshire and Derbyshire border are substantial remains of the headstocks and winding engine house of the Winterbank Colliery which closed about 1887. These were erected half a century or more earlier; and by modern standards they are incredibly small.

The headstocks of a particularly interesting small mine were dismantled in late 1969. These were at Brora Colliery, Sutherland, the only coal mine in the Scottish Highlands. This is a tiny coalfield, remote from the main coalmining areas of Scotland. The earliest workings, near the seashore, were on outcrops; but a pit was sunk by Jane, Countess of Sutherland, as early as 1598, presumably to supply the salt pans with fuel. A second shaft sunk by her son, John, the 5th earl, in 1641 about 200 yards to the north of the old Salt House, is still visible.

The early pits at Brora were shallow: there were some of twelve to thirty yards deep in the eighteenth century. Deep mining at Brora may be said to have commenced in 1810 when a shaft was sunk to a depth of eighty-three yards. The shafts recently abandoned date from this period.

In mid-nineteenth century coal production ceased for a time, but was restarted by the Duke of Sutherland in 1872. The headstocks recently demolished almost certainly belong to this period. At the upcast shaft, the wooden headstocks were originally eighty feet high. The reason for this is that it was feared that the water from the pit would pollute the Brora river if allowed to run away. To avoid this water was raised in a tank to the top of the headstocks, well above pulley height. Here, upon a catch being released automatically, the water was tipped into a larger fixed tank mounted in the headstocks, whence it was piped to the sea about half a mile distant. Some evidence of this earlier use is preserved in the illustration which shows a remnant of the peculiar superstructure. The upcast shaft has been used recently as a subsidiary shaft. It had a small single deck cage big enough, perhaps, for two men, and the pulley wheel appeared from the ground to be no more than two feet six inches in diameter. Coal was wound from the downcast which had steel head-gear. The annual output in recent years has been around 8,000 tons. Most upcast shafts can be readily distinguished because the frame-work of the headstocks is boxed in with wood, brick or concrete so as to prevent the fan from drawing air from the surface. At Brora, however, there was no such lining built round the headstocks. Instead, there was an airtight lid on the shaft top itself.[15]

Perhaps some mention should be made of the Koepe system of winding, invented by Frederick Koepe, a mining engineer employed by Krupps of Germany, although this has never achieved in Britain the popularity which it enjoyed in some parts of Europe. Koepe winding involves the use of an endless rope to which the two cages are fixed. The rope passes over pulleys instead of coiling and uncoil-ing on a drum. The first installation in Britain was designed by Robert Wilson for Bestwood Colliery, Nottinghamshire, and it came into use early in 1883, but was withdrawn a few years later.[16] Probably the earliest Koepe Tower Winder in Britain to prove fully successful was that installed at Ashington, Northumberland, in 1922. This was working until 1969, and at the time of writing is still intact, but is likely to be dismantled soon. Among modern collieries using the Koepe system are Bevercotes and Clipstone in Notting-hamshire.

At some modern mines, coal is now wound in large skips instead of cages; but cages are, of course, still used for manriding.

CHAPTER FOUR

Winning and Working

As we have seen, the two chief systems of winning coal conventionally, were bord-and-pillar (with its variants) and longwall. In the nineteenth century, bord-and-pillar was the normal system in Northumberland and Durham whilst longwall was normal in the Midlands. In Scotland, Lancashire and South Wales both systems were used, but as the century progressed longwall replaced the older system at most mines. In Staffordshire a thick seam subject to spontaneous combustion through oxidation of the coal, and known as the ten yard seam, was worked by a variant of bord-and-pillar usually called 'square work'. Here the workings were divided into squares; and barriers of coal known as 'fire ribs' were left between the squares. Special methods had to be devised, also, for working steeply inclined seams such as are found in some parts of Wales, Lancashire, Scotland and Staffordshire. However, in order to avoid too great an accumulation of detail, it is proposed to deal here with the development of the two main systems.

In early bord-and-pillar, less than half of the coal was extracted. The author of the *Compleat Collier* gives three yards or a little over[1] as the breadth of a bord (or stall) and four yards as the breadth of the intervening pillars. As time passed, the tendency was for the bords to increase in size in proportion to the pillars where conditions would allow. Also, the pillars were often 'robbed' because the men found it easier to hew corners off the pillars than to get coal from the solid face. The pillars were also robbed sometimes by cutting 'jenkins' or 'snickets' through them. In some shallow collieries, both on Tyneside and in South Wales, the bords, four yards or so wide, were separated by pillars as narrow as one yard, but this was not possible in deep mines.

Generally one man worked in a bord although, after 1850, double working was adopted at many collieries. This was considered to be inconvenient because the two men got in each other's way to some extent. They were therefore compensated by an addition to the normal price for the work. The headways, or walls, which usually ran along the main line of cleavage of the coal, were narrow and were therefore worked by one man alone. The bords, cut at right angles to the headways, were much wider, and here two men were set to work with each other. It must be emphasized, however, that double bords were a late development in the north-east and ran counter to the established custom of the district. The author of the *Compleat Collier* for one would never have agreed with double working. For him, accustomed to bords three yards wide, it was 'dangerous for two persons to work together, least [*sic*] they should strike their Coal Pics into one another, or at least hinder one another'.[2]

The bord was the pitman's place between one cavel and the next. The first operation was to undercut the coal, using a hand pick. Next, vertical grooves were cut in the face of the coal and mauls and wedges were then used to break the coal down, taking care to make as little slack as possible. The coal was then filled into corves. Before the Industrial Revolution the shovels for filling, like the mauls and wedges, were made of wood, sometimes with a strengthening band of iron. A typical wooden shovel found in Scotland is 32 inches long, the blade measuring about 6 inches by 6 inches, is slightly rounded.

Explosives were used in some mines in the second half of the eighteenth century but only for stone work and blasting through faults. In the north-east, at any rate, gunpowder was not widely used for getting coal until the 1820s. The shot holes were made by an iron bar with an arc-shaped chisel end or, later in the century, by a hand drill.

In the first half of the nineteenth century, after the shot hole had been drilled, a quantity of loose gunpowder was poured into the hole. Then a needle, slightly longer than the hole, was inserted into the charge. The hole was next filled with clay, the clay being 'stemmed' or 'tamped' firmly. The needle was then withdrawn, leaving a narrow hole in the clay stemming all the way down to the charge. This hole was filled with loose powder, or with powder-filled straws, and a piece of touch paper added. Finally, the touch paper was ignited and the shot-lighter ran as fast as possible. A slow-burning safety fuse was invented by William Bickford in 1831, but many people preferred not to go to the expense of using it.[3]

Gunpowder broke the coal down much more easily than wedges. Its one drawback (safety considerations apart) was that it produced a lot of small coal and slack. Sometimes this had to be riddled out and discarded.

While pillars were left permanently to support the roof in all early bord-and-pillar workings, in many later workings the pillars were removed in a separate operation. First, coal was worked outwards from the pit bottom, leaving rather wider pillars than was necessary to support the roof. This was known as working in 'the whole'. Subsequently the pillars were taken while working back towards the pit bottom and this was known as working 'in the broken'. This ensured that valuable coal reserves were not permanently wasted by being left up to support the roof. It also reduced the proportion of capital expenditure (e.g. on shafts and drains) to total expenditure, since more coal would now be won from a given area.

Unfortunately, working in the broken was unsafe in some mines, causing a 'creep' or 'crush' throughout the workings (i.e. causing convergence of the strata). This was the case at Walker Colliery where all the coal, except for that left in pillars, was exhausted in 1795. Thomas Barnes devised a way for making it safe to work the pillars. He divided the pillars into panels of from ten to twenty acres and built a solid barrier around each panel by stowing. He was then able to work half or more of the pillars inside a panel without fear of the convergence being transmitted outside the barrier. When as much of the coal as was safe had been won, the panel was sealed up to minimize the risk of gas escaping from it.

John Buddle improved on this in 1810. He set out to work on the panel system at a new pit, Wallsend G, from its inception. He made plans which divided the area of coal to be won into panels. In between the panels he planned to leave wide ribs of solid coal. These ribs were to be left permanently to prevent 'creeps'. Within the panels working in the broken was to be practised.

Wallsend G pit was separate from the older Wallsend workings; and the new system proved successful. Buddle claimed that the ribs of coal not only prevented creeps, but would also, in the unfortunate event of an explosion, arrest its progress. Buddle's system bears some resemblance to the much earlier Staffordshire Square Work where wide fire ribs were left between the 'sides of work'. He may, indeed, have drawn on his knowledge of this system in developing panel working.[4]

The long-wall system, to which we must now give our attention, lent itself to the division of labour. Here, instead of one man working alone (or two men working as 'marrers'), there was a team of men at the coal face. In most districts one man did the 'holing' (undercutting of the seam), others hewed the coal, getting it down in large lumps with wooden mauls and wedges (or, later, with hammers and iron wedges). Another man called a rembler in Derbyshire, broke the coal into manageable pieces for the loader to load into corves. The timber was set by a separate shift of punchers (or woodmen) who brought the props—usually called puncheons—into the pit with them. It seems likely that the punchers were in some cases required to cut their own props and not merely to transport them into the pit. This is inferred from the fact that props are sometimes found (for example at Stretton in August 1969) with twigs still attached to them as though they were cut roughly from hedges and trees nearby, and used straight away without being trimmed. This may also account for the twigs which Professor Granville Poole found behind the props in very old longwall workings in South Staffordshire though he believed that their purpose was to hold back the debris.[5] The punchers set a line of props, wedged tightly to the roof with small lids of wood, along the coal face; and they moved this line forward each day, re-using as many of the props and lids as possible. Bars were set only very occasionally where the roof was weak, judging by the evidence revealed in opencast workings; where far more props are found than bars. However, in many cases support was provided by leaving roof coal up. For example, on a coal face exposed at Stainsby Hag in summer 1969 there was eighteen inches of roof coal. Similarly, at Stretton, a foot of roof coal was left up in a seam only three feet six inches thick.

Leaving up roof coal was not peculiar to longwall working. J. C., the author of the *Compleat Collier*, recommended the practice for bord-and-pillar work: 'least [*sic*], by the softness of the Mettle Roof, that Roof should fall down and kill your Miners, or what is also bad, bring a Thrust, or a general Crush in one of your Collieries to close it quite up, and thereby lose the Colliery.'[6]

In Yorkshire there appears to have been less specialization than in other longwall districts in the first half of the nineteenth century. In most cases, two men worked together in a heading and four men in a 'benk', an expression which, in this context, means a section of a longwall face. This anticipates the organization common to the east Midlands in the second half of the century.

The posture of the men on the coal face—whether longwall or bord-and-pillar—depended on the height in which they worked. In some seams, like the thick coal of Staffordshire and most seams in south Yorkshire, the men were able to stand upright to do much of their work. However, whatever the height of the seam, 'holing under' (i.e. undercutting the coal) involved lying down. The cut (called the 'Kerf') had to be made thick at the front of the coal face in order that the holer could attain the full depth of cut required. He might, therefore, begin by kneeling, but as he cut deeper into the face of coal, he would need to lie down on his side and might eventually find himself actually inside the cut, hacking away at the coal further back. Before exposing himself in this way he made sure that wooden supports of some kind were inserted into the kerf. Holing, incidentally, produced a great deal of small coal and slack. The proportion of small coal in thin seam working was consequently rather high. In some circumstances, holing was best done in the middle of the seam rather than at the bottom. For example, in the Main Coal at Moira, Leicestershire, in 1834, there was a band of inferior material eight inches thick about half way up the seam. The practice was therefore to cut out this band with hand picks, and then to lever the top leaf of coal down by driving wedges just under the roof. After clearing this coal, the bottom leaf was similarly levered up. The seam was approximately six feet thick.

Undercutting was, however, much the most common method.

With a seam of medium height, say three to five feet, most of the collier's work was done on his knees. In the thin seams of west Yorkshire, on the other hand, the men had no option but to work lying down. They supported their heads sometimes on 'a wooden board or crutch' according to a witness in 1842. In Northumberland and Durham the colliers similarly used a 'cracket' (i.e. a wooden stool) to support them whilst holing. Two of these, shown in Atkinson's book, belong to the Northern Regional Open Air Museum. One, dating from the eighteenth century, has three legs; the other, from the nineteenth century, comprises a rectangular board with two end pieces. The cracket was placed under the thigh whilst the collier was crouching, or under his shoulder whilst lying down. Another three-legged stool found in opencast workings in north Yorkshire a few years ago indicates a similar practice there.

In the east Midlands, there was less specialization of labour in the second half of the nineteenth century than in the first. The remblers were made unnecessary by the introduction of explosives; and the colliers set their own supports. They sometimes did their own loading, too. There was also, in many mines, a change from the 'big' butty system to the 'little' butty system (where each coal face was divided into sections, called stalls, with a butty, or a pair of butties in partnership, in charge of one stall). This use of the word 'stall' is confusing. It must be distinguished from the stall (or bord) of the bord-and-pillar system.

The creation of 'stalls' along the longwall face, each having its own small team of men organized separately from the rest of the face, militated against specialization of labour. The holer, it is true, continued to specialize as a rule, working for several butties; but the rest of the men were expected to do whatever was required of them. In the most common case, there were two butties employing two or three daywage assistants. Between them they got the coal down and loaded it into tubs or 'dannies' (wheel-less tubs) where the seam was especially thin. In some cases they even did their own holing.

This division of the coal face into stalls, besides acting against the principle of specialization, also created roof control difficulties because some stalls advanced more quickly than others, and it was therefore impossible to maintain a straight face line.

Longwall working by hand, as it was practised in most parts of the country (except the north-east) throughout the second half of the nineteenth century, is the subject of a unique set of photographs taken at Clifton Colliery, Nottinghamshire, in 1895. Some of them are reproduced here. Similar photographs, although not quite so comprehensive, were taken at Brinsley Colliery by the Rev. F. W. Cobb at a slightly later date. Cobb also photographed men working Staffordshire thick coal. There are still a few small semi-anthracite mines working a similar 'tub-stall' system, at Alston, Cumberland.

In Lancashire the evolution of the system of working had a pattern of its own. From bell pits there was the usual progression to heading out into the seam. This may have been the stage reached on

9 (*left*) Horizontal double compound winding engine, Elliott Colliery, East Wales.
10 (*below*) Florence sinking engine, Staffordshire

11 (*above*) Vertical winder, Stafford Colliery, Stoke-on-Trent. 12 (*left*) Vertical winding engine house, Kemball, Stoke-on-Trent. (Now used as an engineering apprentice workshop.)

an estate near Wigan when a plan dating from about 1760 was drawn, because this shows twenty-eight shafts in an area of just over twenty-eight acres. The 'Lancashire system' as it developed in the eighteenth century was a combination of a kind of stall-and-pillar with longwall retreating faces. In the second half of the nineteenth century the traditional forms of bord-and-pillar and longwall were both practised, and the Lancashire system fell into disuse.[7]

Let us turn now to the transportation of coal from the working place to the pit bottom. Scotland was particularly backward in this respect. Even in the early 1840s it was still common for the collier's wife and daughter to carry the coal along the underground roadways and up ladders to the top of the shaft. The baskets, weighing up to 1½ cwt, were lifted on to their backs, a supporting strap passing round their foreheads. Unmarried colliers had to hire unattached women bearers (called 'fremd' bearers) by the day. Bearers were also employed at some of the small pits of the Forest of Dean where the timber lining of the square shafts acted as a ladder.

There are indications that this system was used in several other coalfields at an early date. In Pembrokeshire, for example, an Elizabethan author recorded:

> In former tyme they used not engins for lifting up of the coles out of the pitt, but made their entrance slope, soe as the people carried the coles upon their backes along stayers, which they called land-wayes; whereas nowe they sinke their pitts downe right foure square, about six or seaven foote square, and with a wyndles turned by foure men, they drawe upp the coles.[8]

In Northumberland and Durham the corves were dragged on wooden sledges (sometimes called barrows) or trams to the pit bottom until about the middle of the eighteenth century, one person pulling and another pushing from behind. Wooden railways were then introduced for wagon road haulage at many collieries but by no means all. At a few collieries working thick seams and with a strong floor, horses were used to drag single corves (weighing about five cwt) all the way from the working places to the pit bottom. More often, however, the corves were dragged on sledges by boys (one pulling and the other pushing) to the main road where they were loaded on to wheeled trams, each one holding one corve, to be drawn by horses to the pit bottom.

At about the end of the eighteenth century, the rolley was introduced. This was a wheeled vehicle, like a tram but bigger, capable of holding two or three corves. The corves were still drawn on sledges to the main road as before, but were then lifted by a crane on to the rolley. The rolley was drawn to the pit bottom by a horse. A unique (but technically inaccurate) picture of 'A crane for loading the rollies' is included in T. H. Hair's *Sketches* of 1839. About 1840 square wooden tubs or trams with wheels were introduced at many pits. They ran on iron rails from the working place to the main road, and were then pushed on to rolleys. Soon the absurdity of this was recognized and the tubs were drawn on rails all the way to the pit bottom. At the same time ponies replaced hand putters for bringing the trams to the main road at many pits. A few collieries in this coalfield continued to use hazel corves, however, until the 1880s.

By the 1820s, steam-driven haulage engines were being used underground at a few collieries for hauling trams up dip roads.[9] Long before this, however, self-acting inclined planes (sometimes called 'jigs') had made their contribution to efficiency underground at a number of places. Michael Menzies took out a patent as early as 1750, and in 1800 a self-acting inclined plane, the full wagons drawing up the empty ones, was installed at Townley Colliery near Newcastle.

From about 1850 mechanical haulages on level roads became common in Northumberland and Durham. They were of three main types: main, or direct, rope haulage (where the gradient is just sufficient for the empty tubs to run inbye under the force of gravity, dragging the rope with them), main and tail haulage, and endless rope haulage. With main and tail, the engine is fitted with two drums, one for the main rope which hauls the full train outbye and the other for the tail rope which hauls the empties inbye. The endless rope system requires a double set of rails. The rope runs along the road, round a large wheel at the far end and back to the drum. The trams are usually clipped to it in sets.

Before we discuss motive power, which is common to all districts, perhaps it would be best to survey briefly the development of underground haulage in other coalfields in the period prior to 1850.

In most districts the corf was used to convey coal. An exception was in the small Radstock (Somerset) coalfield where the coal was loaded into wooden sledges having wicker or wooden sides; but even here, it was transferred into corves in the pit bottom for winding up the shaft. In shallow seams the Somerset sledges were drawn by children on hands and knees. The hempen harness was called the 'guss'. This continued in use into the twentieth century, and there are still people alive who wore the guss as young men. Indeed, the example at the Lound Hall Mining Training Centre was worn by the person from whom it was purchased. There was also a late survival of drawing by leather dog belt at the Jackfield pits, Madeley, Shropshire. A. Yates, who died in 1969, was so employed in the early years of this century.

In Yorkshire, where there was sufficient headroom, wheeled trams had replaced the wheel-less corves by 1841. This innovation was the work of John Curr, who introduced such trams into the Sheffield district about 1790. To hold the trams steady in the shafts he used guide rods (patented in 1788) which were deal rods of four inch by three inch section running down the side of the shaft, forming mortices. The trams were suspended on wooden cross-bars with rollers at each end. The rollers ran in the guide rods. When a tram reached the surface it was lifted a little above ground level to allow a wooden platform to be inserted beneath it. The tram was then pushed off the platform and wheeled away on rails. Unfortunately these trams or tubs were still referred to by John Curr as 'corves' and this name continued in use for a long time after.

While the use of wheeled trams spread to many other Yorkshire collieries, the guide rods did not, and there were still many shafts without them in mid-century. The principle of shaft guide rods was sound, but in practice John Curr's arrangement was by no means wholly successful. Not until the invention of the cage did guide rods win general acceptance.

Before the introduction of rails in the second half of the eighteenth century the most usual method of underground transportation was to drag corves, usually mounted on wooden runners or sledges, to the pit bottom. Sledges have been found in old workings in most parts of the country.

Even after the introduction of rails on the main roads, which we have discussed elsewhere, the corves were still usually dragged to the main road, without the aid of wheels (except in Yorkshire), and were then loaded on to wheeled trams. The following description of the system in the Nottinghamshire and Derbyshire coalfield was recorded by a Children's Employment Sub-Commissioner in 1842:[10]

> At seven years old the boys drive between—that is, the corve without wheels, with from eight cwt to nearly a ton is drawn the length of the bank (mostly about 200 yards) by three asses. 'The between driver' is placed behind the second ass and has to attend to the two first—the last is driven by the ass-lad who is often not more than 12 years of age. The elder boy wears a dog-belt, but not to draw with continuously, the descent frequently being sufficient, or even more than sufficient, for the corve to run without much drawing; the elder boy walks backwards, and has at the same time to urge the last ass on, and by his belt prevent the corve running against the side of the bank; when the corve reaches the waggon-road it is placed on wheels and left to the care of two other boys (one perhaps about 13, the other 8 or 9 years old;) the elder one wears the dog-belt and occasionally draws by it, or in some pits where the descent is good, he merely uses the 'crop-stick'; in returning the youngest boy goes before the waggon and the elder pushes behind; until a boy gets accustomed to the dog-belt it frequently produces soreness on the hips and otherwise injures him.

Asses were used instead of ponies because the roadways (except the main road) were driven in coal only and were usually not high enough to accommodate ponies. Even some of the main roads were driven the height of the seam. In many cases, the headroom was between three and four feet, and both boys and asses injured themselves constantly by rubbing against the roof.

In some cases, where very thin seams were worked, there was not sufficient height even for asses. Here, the corves were drawn (usually on sledges) by boys crawling on all fours. In some districts (but not Nottinghamshire and Derbyshire) girls were also used for this work. At some collieries, the corves were pushed, but more often they were pulled by means of a dog-belt (a broad leather belt) fastened round the waist or hips, to which was attached a chain which hooked on to the corve. A similar method was, as we have noted, still used at some Somerset collieries at the beginning of the twentieth century, although the guss was worn there instead of the dog-belt.

The guss harness was also worn in Lancashire in the eighteenth and early nineteenth centuries by both boys and girls, and also by women. One such harness, made of hemp rope, was found at Atherton and was deposited at the Wigan Mining and Technical College. As elsewhere, the hazel corves were placed on sledges (here fitted with iron runners) and since the seams, which determined the height of the roadways, were not very thick the hauliers worked on hands and knees. Anderson tells us that the corves used at the Hulton pits near Bolton measured thirty-one inches by twenty-one and three-quarter inches by eight inches deep and they held about 1½ cwt of coal.

Sometimes, wooden rails were laid for the sledges to run on. Some found at Orrell near Wigan forty years ago were in five-foot lengths. They were five inches wide and one inch thick. A one-inch square strip of wood was nailed along one edge to guide the sledges.

By 1840 many collieries in Lancashire were using wheeled sledges on which the corves were placed. The plain wheels ran on rails made of angle iron. This system gave way to cages and tubs with flanged wheels in the 1850s, Lancashire being on the whole rather backward in this respect. For example, the Bye pit at Pemberton (near Wigan) changed over to the new system on September 11th 1855 using tubs of six cwt capacity; and the Venture pit belonging to the same group changed over some three years later with tubs of five cwt capacity.[11]

In Yorkshire, as we have seen, wheeled trams were used in many collieries at the time of the Children's Employment Commission, 1842. In the Halifax district, for example, trams holding from two to five cwt and having four cast iron wheels about five inches in diameter were hauled by children. On the gates leading to the coal faces the height was in some cases as low as sixteen to twenty inches. The children dragged the trams, using the dog-belt, to the main road. Then they were able to push the trams on the rails to the pit bottom. Older children worked alone on this task, but young ones worked in pairs. In this county, as in Lancashire, girls were employed underground as well as boys until 1843 when it was made illegal to employ females, or boys under ten years of age, underground.

In 1842 there were some pits with thin seams where all the roadways were seam height only. It was impossible for men to work in some of them at all, and even children were unable to stand. Thus at Brampton near Chesterfield the roadways were only two feet high, and the only place where a boy could stand was at the bottom of the shaft. Here the corves were of necessity drawn by boys to the pit bottom. The average corve load was one cwt and at one pit the boys made sixty journeys a day, over a distance of sixty yards. Similarly, in west Yorkshire, very thin seams were worked. In the Flockton collieries the main roads were from twenty-two to thirty-six inches high, and the side gates were lower still.

In workings forty to fifty feet deep exposed in the high wall of an opencast site at Stretton, Derbyshire, to which we have referred earlier, the roadways were only about two feet six inches high, driven under about a foot of top coal. Of course underground transport in these circumstances would be expensive, and shafts were therefore sunk at frequent intervals so as to keep the cost down to a minimum. In the Stretton case, four shafts could be seen in the high wall in a distance of about 180 yards, one being 'dog-legged'. It seems that the dog-legged one was sunk vertically for about twenty-five feet, when a landing was created and a further vertical section sunk down to the seam. This was doubtless to accommodate ladders for travelling the shaft. However, the coal was presumably wound from the vertical shafts and not carried up the ladders. This corresponds with the system at Smalley, near Derby, in 1693, when a visitor reported that he went down a mine 120 feet deep on ladders of twelve staves each 'set across the pit one by another', but where the coals were drawn by a horse gin.[12]

Even where height permitted the use of asses or ponies, the owners of some small collieries did not employ them but instead relied on women and children.

The wide adoption of shaft guides and cages in the 1840s and 50s was mainly due to the fact that they increased the efficiency of the winding operation and therefore the profitability of the pit. So did the adoption of wheeled trams and roadways high enough to be travelled by ponies. The tram, hauled by ponies the whole way from coal face to pit bottom on rails and then pushed on to the cage, marked a great advance in mining technology. Now the coal could complete the whole circuit to the screening plant or landsale wharf in trams drawn on rails.

In thick seams like the Top Hard or Barnsley bed of the Yorkshire, Derbyshire and Nottinghamshire coalfields or the Staffordshire Ten Yard Coal, it was possible to employ horses rather than ponies; and many pits did so in the second half of the nineteenth century. There are several photographs of underground stables with shire horses in the author's possession.

In the Whitehaven district, pack-horses were used in the late eighteenth century to carry coal out of drift mines. According to C. Humphreys, some of these drifts were T-shaped in cross-section, presumably to provide ledges for the baskets to rest on. The mouth of one drift (locally called a 'bearmouth'), at Howgill, Whitehaven, can still be seen.

The William pit at Whitehaven was one of the last to use corves. These were not abandoned until 1875. There are a few surviving Whitehaven corves, and one has been reserved for the Lound Hall Mining Museum.

To return now to mechanical haulages, it should not be thought that these were peculiar to Northumberland and Durham. On the contrary, even in Lancashire and Cheshire stationary steam engines were being used at several mines for hauling coal from the dip in the 1820s. Many other coalfields had a few such engines at work underground in the 1830s. At Fenton Park Colliery, Staffordshire, for example, coal was being hauled mechanically up five inclines, one of them 400 yards long, in 1839; and in 1841 there was even a haulage drawing coal up a steep incline to the pit bottom at Newfield pit, Moira, in the South Derbyshire coalfield which was generally in a rather backward state. At one of the other Moira pits, Bath, a four-horse gin was similarly used to draw trams up an incline, one of the few recorded instances of the use of horse gins for underground transport. Probably the first colliery to have a mechanical haulage operating on a level road was Peacock Mine, Hyde, where one was working in 1841.[13]

Again, the pioneer work on wheeled tubs (or trams) was done not in the north-east but at Sheffield by John Curr.

By 1850 the danger of underground steam engines was causing concern. In some places, this was overcome by putting the engine on the surface and running the rope over a pulley and down the shaft. A remarkable example of this was the 'Shonky' engine at Bulwell Colliery (closed 1945), an old railway engine mounted upside down with its flanged wheels in the air, which drew 700 tons of coal a day from the coal faces to the pit bottom. When this engine was going flat out it danced up and down on its bed in a manner which visitors found disconcerting. There was a similar example at Hanwood Colliery, Shropshire, until 1940. At Alfreton Colliery, Derbyshire, there was a portable engine supplying power for the underground haulage at the turn of the century. This engine had two cylinders of ten inch diameter, sixteen inch stroke, revolutions per minute about 130; it drove on steel wire rope three-quarters of an inch in diameter going down the upcast shaft over a pulley to drive the gearing of an endless rope haulage in the pit bottom. This engine was capable of winding the men up the upcast shaft should there be an accident in the downcast.

An alternative method used, for example, at Donisthorpe Colliery, South Derbyshire, at the close of the nineteenth century and in the early years of the twentieth, was to convey steam down the shaft in pipes to operate a haulage engine in the pit bottom. A report on Donisthorpe in 1903 said that this system was inefficient on account of the loss of steam and condensation in the pipes; and this was the general experience.

At the Pemberton Collieries in Lancashire, the late nineteenth- and early twentieth-century haulages were driven in some cases by engines situated on the pit top and in others by compressed air engines underground. Typical of the former was the Orrell Five Feet engine. This was a double cylinder engine, the cylinders being of sixteen inches diameter and thirty-inch stroke. Two three-quarter inch ropes passed over pulleys set in a frame in the headgear down the 580-yard deep shaft through five inch square wooden pipes, passing round two more pulleys in the pit bottom.

The tram track at this colliery was of two foot gauge (perhaps the most widely adopted gauge for tub circuits) and the rails were originally eighteen pounds per yard in weight, but heavier (twenty-five pound) rails later became standard. On the main and tail haulages here, tubs ran in sets of forty to sixty.[14]

Another solution was the adoption of compressed air. The first British colliery to use this was probably Govan Colliery, near Glasgow, where a compressor was installed for haulage engines in 1849. Progress with compressed air was rather slow, and in 1883 it had a new competitor. In that year, an underground endless-rope haulage driven by electricity was installed at Nostell Colliery, Yorkshire. The first alternating current (AC) haulage engine was installed at Denaby Main, Yorkshire, in 1900.[15]

Compressed air and electricity, besides superseding underground steam engines, facilitated the widespread use of mechanized haulage systems. Even so there were still over 65,000 pit ponies and horses at work underground in 1924.

In modern mining practice, locomotives hauling large mine cars, and conveyer belts have largely taken the place of earlier underground transport systems; but the transport of materials inbye has not kept pace with the transport of coal outbye and there is still considerable room for improvement.

There are now very few collieries employing ponies underground and by the time this book is published there will be hardly any.

Ventilation

Ventilation was no problem with bell pits and shallow drifts. Once this stage of mining was passed, however, difficulties appeared. In a mine with only one shaft there will be a natural flow of air only so long as the temperature below ground is markedly higher than that on the surface. Warm air rises, and in winter a convection current would be set up in the shaft sufficient to ventilate a small shallow pit. In the summer, however, there might be little flow of air or none at all. In these circumstances the men were in danger of being overcome by the gas we know now as blackdamp or chokedamp. In the sixteenth century they called it 'styth' or 'dampe'. It is a mixture of carbon dioxide and nitrogen in varying proportions and it extinguishes life. It may be either lighter or heavier than air, depending on the proportions of its two constituents.

According to Roger North, writing of the Northern Coalfield in 1676, miners in the summer 'tried' the air by lowering a dog down the shaft. If he howled it was unsafe to go to work. North also referred to the effect of the gas on the men's candles. The lowering of the flame was a sure indication that gas was present.

In the sixteenth century and earlier, when styth was found to be present, the men tried to beat it out of their working places with a jacket. They knew of no other remedy. By Roger North's time, however, it was recognized that two shafts interconnected (or a shaft and an adit) provided a flow of air and reduced the risk of suffocation from chokedamp.

In everyday terms warm air rises because it is lighter than cold air. Natural ventilation results from the temperature underground being normally greater than the surface temperature. Below a certain depth, temperature rises by about one degree Fahrenheit for each sixty feet, and to this must be added the heat given off by men, animals (and, nowadays, machinery) working below ground. Ventilation will be better in the winter when the surface temperature is low than in the summer when it will approximate more closely to the underground temperature.

Where two shafts are of unequal depth, in winter air will flow down the shallow shaft and up the deep one, because the temperature in the shallow shaft will be lower than that in the deep shaft. In summer the temperature in the shallow shaft will usually be higher than that in the deep shaft and so the direction of the air current will be reversed. With two shafts of equal depths, the direction of the air current may be determined by the direction of the prevailing wind, or by one shaft being wetter (and therefore cooler) than the other. Indeed, one of the methods sometimes adopted to help ventilation was to pour cold water down the downcast shaft. In some mines, instead of two shafts, there were one shaft and one adit. A fluorspar mine at Crich, Derbyshire, still relies on natural ventilation from an old sough. In some fair-sized drift mines, in the Forest of Dean particularly, a floor of wooden boards was built leaving a small channel underneath the boards for drainage. This also acted as the air intake, the stale air returning to the surface through the upper section of the drift.

Many two-shaft collieries were still badly ventilated and gases accumulated in them, especially when the temperatures (and therefore the weights) of the two columns of air were about equal and there was therefore hardly any flow of air. This was often the case in spring and autumn. It was also noticed that the direction of the wind could affect ventilation considerably at some collieries.

In the seventeenth century, it became recognized that ventilation could be assisted by the use of a basket of fire in one shaft which then became the upcast. The first recorded use of a fire basket in England was at a colliery at Cheadle in north Staffordshire, in the mid-seventeenth century.[1] However, this practice spread very slowly. The first recorded example in the north-east was at Fatfield Colliery in 1732; and in the Midlands there were still collieries relying on purely natural ventilation a century later. The owners of fiery collieries had a greater inducement to ensure efficient ventilation than others.

This brings us to a consideration of the second main gas met with in collieries: firedamp, which the men often called 'wildfire'. Firedamp (or methane) is created by the decay of vegetation: it is, indeed, identical with marsh gas. The firedamp found in pits resulted from the decay of vegetation many thousands of years ago when the coal seams were being formed. It is retained, under pressure, in the seam and is released when the coal is worked. Most shallow mines are free from firedamp and it was, therefore, of little importance until the seventeenth century. By the opening of the nineteenth century it was a scourge in the deep pits of Northumberland and Durham. According to Galloway, there were 643 colliery explosions between 1835 and 1850, and some of these killed a great many people. It also damaged the colliery owners' property. Thus an explosion at Fatfield in 1708 destroyed the winding gear besides killing sixty-nine miners. Again, at mines near the Mendip Hills in the late seventeenth century, the force of explosions threw the winding gear off its frame, and did other damage to the shafts, ropes and so on. To minimize the risk of explosions the men in this district used small candles which they kept behind them: they did not 'present them to the breast of the work'.[2]

Miners soon learnt that the presence of firedamp could be detected by its effect on the candle flame. Whenever the gas was present it created a blue 'cap' above the normal flame. In fiery mines it was the practice to test for gas by gently raising the candle up towards the roof (since methane is lighter than air) and into cavities. A blue cap of more than a certain size called for speedy withdrawal.

But it was not possible to test for gas the whole day long and explosions took place at the candle flame without warning. Many of these were minor affairs causing no loss of life, but men working in fiery mines expected frequent burns. In the Mendip case the colliery owners kept ointment handy for the treatment of such burns 'being furnished therewith at the cheap rate of 12d. the pound, by a good old woman living near the works'.

The best way to avoid methane explosions is to have an efficient system of ventilation so as to dilute the gas: explosions only occur when there is between about 5·6 and 13 per cent of methane in the air. Where the flow of air was insufficient to dilute the gas then the only way to make the workings safe was to ignite the gas deliberately. This was done at Mostyn Colliery in North Wales as early as 1677. Here a man (who came to be called the 'fireman') entered the workings at the beginning of the shift, covered in old rags soaked in water. He carried a long pole, to the end of which he attached a lighted candle. He thrust this into cavities in the roof, meantime lying close to the floor face downwards, until the methane took fire. The workings were then considered safe for the shift. During the eighteenth century, this and similar methods of firing the gas became widely adopted. Because of the rags he wore for protection, the fireman in some districts was called 'the penitent'.[3]

Men working in the deeper pits, of Northumberland and Durham in particular, were always at risk; and considerable attention was given to the possibility of adopting some safe means of illumination. For working in the shaft or pit bottom area, reflecting mirrors were sometimes used. The first safety light for use inbye was the Flint and Steel Mill invented by Carlyle Spedding of Whitehaven about the middle of the eighteenth century. Here, a steel toothed wheel was made to rotate in contact with a flint thus emitting a continuous stream of sparks. It was expensive in use, because every collier needed someone with him to operate the mill, and it was expensive to keep in repair. Further, it provided very little real protection because firedamp can be ignited just as easily by a spark as by a flame.

Some miners tried working by the light of decaying fish and other phosphorescent materials but this cut down their productivity besides ruining their eyesight.

Credit for the invention of a trustworthy and efficient safety light is usually given to Sir Humphry Davy. His lamps were based on the principle that if the flame is surrounded by wire gauze of a fine mesh, it will not cause an explosion. Prototype Davy lamps were tested by Matthias Dunn and the Rev. John Hodgson at Hebburn Colliery, a particularly fiery mine, on 9 January 1816. The tests were completely successful (Fig. 11).

Two other men, Dr Clanny and George Stephenson, also invented safety lamps at about the same time and quite a number of other people deserve credit for improving on the earliest lamps. For example, George Upton and John Roberts invented an improved lamp in 1827.[4] Subsequent improvements were designed to increase the light output of the safety lamp. This was achieved by having a glass cylinder round the flame itself and a double gauze mounted in a cylindrical casing above the flame (Fig. 12).

Miners complained that the safety lamp was a mixed blessing because some colliery owners expected men to work with it in places full of gas where they could not have worked with candles or steel mills. This complaint was echoed in the middle of the nineteenth century by mines inspectors who said that the lamp was being used as a substitute for good ventilation.[5] Here it should be remembered that by this time the use of gunpowder was almost universal, and that this was a major risk in a gassy mine, whether safety lamps were in use or not.

There is no real substitute for good ventilation. We saw earlier that firebaskets came into use at some places in the Midlands in the seventeenth century, and in the Great Northern Coalfield eighty years or so later. The fire basket improved the rate of flow of air, but not necessarily its circulation round the workings. The method of circulating the air in large mines before about 1760, was known

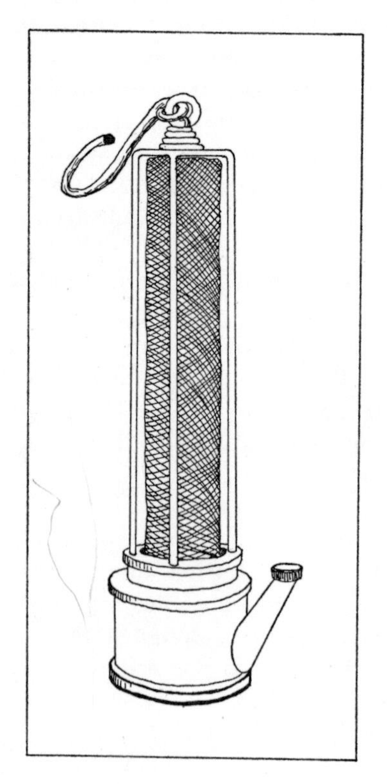

Fig. 11
Original Davy lamp

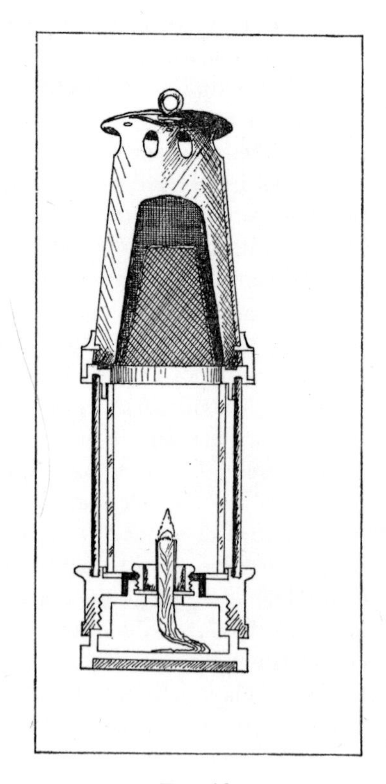

Fig. 12
Modern Marsaut-type safety lamp

as 'face airing'. The air went down the downcast and was then directed by means of stoppings round the working faces leaving gas to accumulate in subsidiary roadways. James Spedding initiated an improved circulation, called 'coursing the air', about 1760. This ensured, by the use of doors and stoppings, that the air traversed all the roadways as well as the faces. The drawback of this system was that in a very large colliery working bord-and-pillar the air might have to travel twenty or thirty miles before reaching the upcast by which time it would be both sluggish and heavily polluted. Also, a door left carelessly open could short circuit the flow of air, putting half of the mine at risk.

In the first decade of the nineteenth century, 'air coursing' gave way to a new method by which the main air current was split so as to provide all the faces with fresh air. This system also minimized the harm done by leaving a door open.[6]

The necessity of directing the flow of air by doors had the unfortunate effect of providing work for very young children. It was usual for door 'trappers'—girls as well as boys in some districts—to start work at six or seven years of age. The trapper's duty was to sit by a door all day long, closing it every time someone passed through. George Parkinson of Durham described his experience when starting work as a boy of nine in 1837, as follows:

> Everything was new and strange to me; and as we passed along the narrow wagon-way, with its wall of coal on either side and its stone roof so near, it seemed to me a little world to live in. A few hundred yards brought us to a large trap-door about six feet square, closing the whole avenue. This was to be my abiding place for the next twelve or thirteen hours, and my father set to work to make a trapper's hole behind the props, in which I might sit safely and comfortably. After hewing out a good shelter for me he put a nail in the door, to which he fastened my door-string, attaching the other end of it to a nail in a prop where I sat, so that I could pull the door open when the horse and wagons were coming through without exposing myself to danger. Then, after showing me where to stick my 'lowe', i.e. to place my candle, and giving me full instructions, he went away to his own work, and I was left alone.

A little later, a colliery official came by:

> He looked very sternly at me as he held up his stick in a threatening way, and said 'Now mind, ef thoo gans to sleep and dizzent keep that door shut, thou'll get it.'[7]

By this time, collieries in the north of England were using ventilation furnaces, and not fire baskets. These were usually, though not invariably, built underground. The first recorded use of a furnace in this country was, indeed, a surface installation. This was erected at a Wearside colliery, North Biddick, in 1756, and was similar to the so-called 'air tubes' used at Liège in the mid-seventeenth century. A furnace was built near the top of the upcast shaft and was surmounted by a tall chimney, the whole being completely enclosed by brickwork so as to prevent air from reaching the fire from the surface.

Perhaps the first underground furnace was the one installed at Wallsend in 1787. This was so clearly superior to surface furnaces that most other collieries in the north-east followed suit. At Wallsend, and many other mines, a brick chimney was built over the upcast. This had an aperture at the top for releasing the smoke and foul air into the atmosphere and the whole of this top part of the structure revolved with changes in the direction of the wind by the agency of a wind-vane. This prevented the wind from blowing through the aperture, thus avoiding the risk of having smoke forced into the workings and the flow of air impeded. At first the return air from the workings passed directly over the fires; and explosions sometimes occurred. A subsequent innovation lessened the risk of explosions. This was the dumb drift which isolated the return air from the furnace but it was not universally adopted.

There was also a risk, even if not a particularly great one, that the furnace would start a fire in the pit bottom area. This happened at the St Hilda Colliery, South Shields, during the night of 14 January 1841 when the flames spread to the wooden brattice in the shaft. Like many northern mines, St Hilda had only one shaft, divided vertically by a brattice to form a separate upcast and downcast. It was feared that the destruction of this brattice would stop the circulation of air in the pit, thus putting at risk the lives of the twenty or so men and boys who were at work underground. Fortunately the water poured down the shaft to extinguish the flames had the additional effect of assisting the downcast flow of air. The

13 Old roadway, 3 ft 2 in high, Brinsley Colliery

14 Flue of ventilation furnace at Brinsley

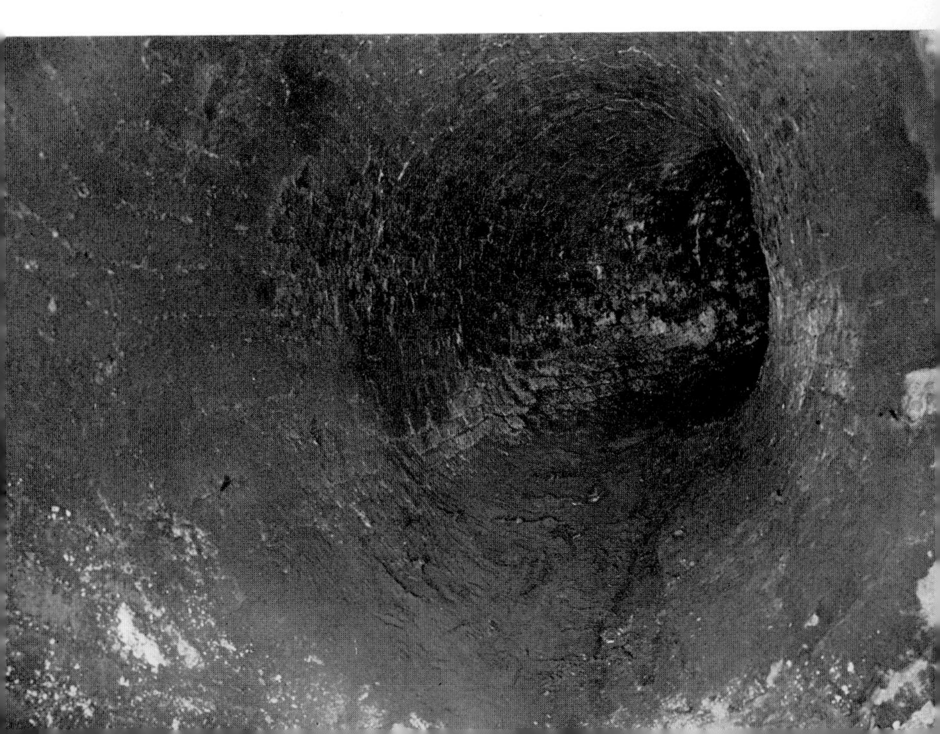

15 Holing under at Brinsley, 1910–13

16 Brinsley screens, 1910–13

fire was extinguished and the miners were brought safely to bank. At another Tyneside colliery, Ellison Main, five men were suffocated on 11 December 1815 as a result of the brattice taking fire from a fire lamp in the shaft bottom, thereby stopping the ventilation of the workings. A fire lamp in the pit bottom was, of course, more dangerous than a proper furnace a little way out of the bottom.

Furnace ventilation was a great improvement, but even so, for many years the flow of air was still grossly inadequate. The result of this may be exemplified by another passage from Parkinson's autobiography relating to the late 1830s:

> On another occasion two other lads were working with me at night, our business being to fill loose coal into tubs and bring these to a place appointed. No other persons were working near us. The mines were not so well ventilated then as they are now, and having learned by experience the danger that existed where the current was weak, I took extra precautions. My friends were working a little distance beyond me. The feeble current of air, with nothing to direct it, was passing from me to them. I fixed my candle about two feet from the floor of the mine at the end of the place and went up a few yards where it was almost dark, to fill a tub. My mates came along with me and one of them suddenly exclaimed 'Parkinson, what's the matter wi' thi' lowe? It's blue up the sides.' Looking round, I perceived instantly that the place was filling with gas, and that an explosion was imminent. I said in a whisper 'Be perfectly still; don't move a hand.' I lay down and cautiously crept towards my candle. I was afraid to breathe, for the slightest movement of the flame might cause the gas to ignite. Slowly I raised my hand to the candle. . . I drew the light down very gently and slowly inverted it so that the melted grease should extinguish the dangerous flame.

Nevertheless, a sufficient flow of air could be induced by fire. This was achieved by building larger roadways, splitting the air current, and having two or three large furnaces instead of one small one. In this way Hetton Colliery in 1835 was provided with 93,300 cubic feet of air per minute against the 5,000 or so cubic feet common at most large collieries in the north. In the 1840s standards of ventilation in Northumberland and Durham improved immeasurably.

Elsewhere very few collieries were well ventilated in 1835.

Lancashire collieries had only very recently gone over to underground furnaces, and stoppings there were usually heaps of debris instead of solid walls so that the circulation of air was not properly controlled. A few of the deep mines in north Staffordshire had furnace ventilation, but most mines in this and adjoining counties relied on natural ventilation, helped, here and there, with a fire basket. This was still true of Nottinghamshire and Derbyshire some seven years later, except for one or two isolated examples of furnaces as at George Stephenson's newly opened Clay Cross Colliery. This is rather surprising because, according to Pilkington's oft-quoted *View of the Present State of Derbyshire* most Derbyshire mines had a form of furnace ventilation in 1789. In Pilkington's words:

> At most works there is besides the large shaft, by which the coals are drawn up, a smaller one at a distance of a few yards. This is about 4 ft wide, and 15 ft or 16 ft deep, and from the bottom of it a pipe is carried into and down the larger shaft to that part of the mine where the men are at work. A vessel of burning coal, holding about four pecks is then suspended in the smaller shaft. By this contrivance the air is immediately rarefied and a fresh column rushing upwards to supply its place, a circulation is produced and maintained in every part of the mine.

No doubt Pilkington saw such an installation but his view that this system of ventilation was general in the county is untenable.[8]

There were few mines in South Wales making use of artificial ventilation, largely because many of them were drifts which do not lend themselves readily to furnace ventilation.[9]

Many of the small collieries of the Midlands which relied still on natural ventilation, had not merely two, but several shafts. This was partly in order to minimize the cost of underground transport and partly to provide a flow of air. Whilst many of them were described by witnesses to the Children's Employment Commission as 'well-winded', it was acknowledged that they were affected by blackdamp, particularly in summer or when the wind was in a particular direction. Some boys said that the gas made their bellies swell and their heads ache; and it undoubtedly caused respiratory weakness. Most middle-aged miners in the Midland coalfields at this period suffered in consequence from 'miners' asthma', characterized by shortness of breath.

From mid-century, even in the Midlands, efficient furnace ventilation with good roadways became common except at the very small mines (especially in Staffordshire and Shropshire) which were run by subcontractors. One of these furnace systems, installed at Walsall Wood Colliery, South Staffordshire in 1879, was in use until 1950. This furnace had a grate six feet nine inches by four feet wide. It was built of bricks and lined with firebricks. In 1925 its capacity was 100,000 cubic feet of air per minute, delivered at an average cost of 1·16 pence. The colliery was therefore ventilated for less than £7 per day.

A scale model of the Walsall Wood furnace was exhibited at the Wembley exhibition in 1925. Unfortunately the buildings still standing at Walsall Wood (which closed about 1964) are unrelated to the system of ventilation, so no direct archaeological evidence remains.

Ventilation furnaces were, as we have seen, sometimes built on the surface and one of these was in use at Broseley Deep Pit, Shropshire, until 1941. Perhaps the last fire basket to be used was also at a Shropshire mine, the Rock pit, in 1965, but only at week-ends when there were no men underground. This was photographed by Ivor Brown.

Quite a number of old collieries which were originally ventilated by furnace are still open. In most cases the furnaces were on levels which are now inaccessible, but there are a few where the site of the furnace can still be seen. One such is Brinsley Colliery, near Eastwood. Here the flues and furnace housing are still to be seen at the time of writing although the furnace went out of use well back in the nineteenth century. Brinsley became an upcast for a neighbouring mine and the fan, fan drift and housing belonging to this period are still to be seen. However, the engine was removed about 1965 and both shafts then became downcasts. An explosion, mentioned elsewhere in this account, occurred in 1883 in the Brinsley pit bottom owing to the fact that the fan was allowed to stand at week-ends for reasons of economy. In Lancashire ventilation furnace chimneys can still be seen at Clifton Colliery, Burnley, and Pewfall Colliery, Garswood. A surface ventilation furnace and chimney can still be seen at a drift mine at Trehafod, in South Wales.

If the Midland coalfields lagged behind the north in regard to ventilation furnaces, at least they had few single-shaft collieries. As we saw in an earlier chapter, mining engineers in Northumberland

and Durham, faced with the cost of sinking to great depths, economized by making one shaft do the work of two. This they did by dividing the shaft vertically with wooden brattices into two (or sometimes three or even four) sections. One section was connected with the underground ventilation furnace and solid stoppings were used in the pit bottom to make sure that the flow of air was not short-circuited. Thus the air went down the downcast section of the shaft, through the workings, then on its return journey passed over the furnace (via the dumb drift) and up the upcast section of the shaft.

It will be apparent that this system depended on the reliability of the wooden brattices used to divide the shaft. If these were destroyed by fire, or explosion, or other physical calamity, then the flow of air stopped.

The danger may perhaps be best exemplified by considering further the Hartley Colliery disaster of 1862. This pit was 600 feet deep and much affected by water. The water was raised in three stages at the rate of 1,500 gallons per minute, the motive power being provided by an engine made by Losh, Wilson and Bell which had a cast iron beam weighing forty-two tons. This beam operated the pump-rods in the shaft. On the day of the tragedy, the beam broke in two and one half of it crashed down the shaft, ripping out the brattice and much of the flimsy wooden shaft lining. The shaft was completely blocked with debris.

There were three seams at Hartley in which there were roadways; the High Main first, then the Yard Seam and then the bottom seam, the Low Main, in which the men were working. There was a staple shaft (a shaft connecting two levels below ground) which enabled the men to get up to the Yard Seam. There was also a separate pumping shaft connecting the High Main with the surface. There was, however, no interconnection other than the main shaft between the Yard and High Main levels, a distance of sixty-six yards, and it was this portion of the shaft which was blocked by debris. The men were unable to proceed any higher than the Yard Seam and there they waited while rescue teams cleared the debris. At first the real danger was not fully apprehended. This was that without the brattice there was no ventilation. Unfortunately, before the shaft could be cleared, the men and boys were overcome by blackdamp. Altogether, this disaster cost 204 lives. All that remains to be seen of the old Hartley pit is the square brick wall surrounding the main shaft (into which is set the date stone from the colliery engine house) and a similar wall round the adjacent pumping shaft.

As a direct consequence of the Hartley disaster, the law was amended to make it compulsory for every colliery to have at least two shafts.[10]

By the time of the Hartley disaster, furnace ventilation had started to give way to mechanical fans. In early mining practice, small hand-driven fans (and, before that, bellows) had been used to force air through wooden pipes to ventilate headings. In the 1840s, such fans were still being used in the Midlands and south of England to ventilate headings and shaft sinkings. They were used, for example, for shaft sinking at Babbington, near Ilkeston. The most popular hand-driven fan was known as the 'blow-george'. Its usefulness was very limited. It was no good for deep sinkings, where small furnaces and temporary brattices to divide the shaft into downcast and upcast were used instead. The only 'blow george' now known to remain *in situ* is at a derelict lead mine in Shropshire. I. J. Brown took drawings and photographs of this recently.

In the first half of the nineteenth century, many different mechanical ventilators were invented. Most of the early ones were air pumps. For example, a temporary ventilator introduced by John Buddle at Hebburn Colliery in 1807 was a steam-driven air pump having a five-foot square piston made of wood, working in a wood-lined cylinder at a rate of twenty strokes a minute. This was placed at the top of the upcast shaft and it extracted mine air at the rate of 6,000 cubic feet a minute. Some of the other air pumps invented forced air into the mine. Such ventilators were much less effective than large furnaces.

Besides the air pumps, there were several types of fans, either extracting air from the mine or forcing air into it. Some were barrel-shaped, others were wheels, some were mounted horizontally and others vertically, but they all had vanes which rotated at speed. Compared with modern fans, the speed of rotation was low. For example, a fan invented by William Furness of Leeds in 1837 rotated at a speed of 100 to 300 revolutions per minute and most other fans of the period were much slower than this whereas speeds of 400 r.p.m. are common today.

In a work of this kind it would be inappropriate to describe all the early mechanical fans. One of the most widely adopted in the 1860s and 70s was the Waddle fan, and perhaps this may be taken as typical of its period. It revolved slowly: at about seventy revolutions per minute. It was an exhausting fan, that is, it drew air from the mine up the upcast shaft. It had vanes which were curved backwards

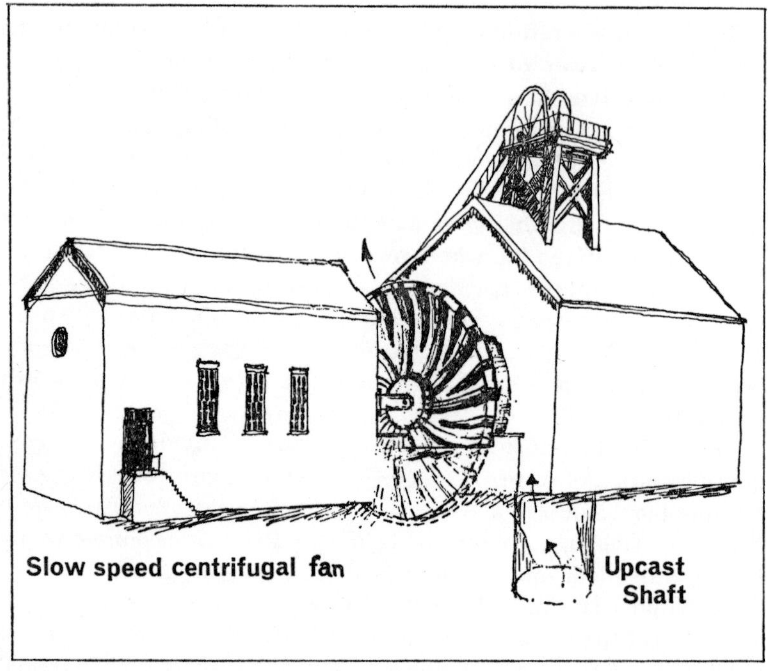

Fig. 13 A waddle fan layout

but, unlike the vanes in some other fans, they were not enclosed in an outer casing. Air entered the fan through a curved passage (Fig. 13). Between 1871 and 1896, 220 Waddle fans were installed. In size they ranged between nine feet and forty-five feet in diameter. In 1947 there were still quite a number of Waddle fans in use, but few now remain. There is one at Annesley, Nottinghamshire and another at Ryhope Colliery, Durham (where coal production ceased recently). This has a diameter of thirty feet and operates at about 100 r.p.m. At the time of writing, there is another Waddle fan about seventy-five years old still *in situ* at Pegswood Colliery, Northumberland. This colliery closed in 1968, so the fan is in danger. Another popular fan of the period was the Guibal. One of these made by

Walker of Wigan was in use at Alfreton Colliery, Derbyshire until well into the twentieth century. It was twenty-four feet by eight feet, and was driven by a high pressure compound engine (with a condenser), having a cylinder eighteen inches in diameter and two foot stroke. This fan worked at fifty-five r.p.m. to produce 69,745 cubic feet of air per minute in 1898. This would have been quite inadequate for a large mine, but Alfreton was of a modest size and its shaft only 152 yards deep.

Modern centrifugal fans rotate at much higher speeds and provide a much greater flow of air, but the principle is still the same. The fan is a horizontally mounted vertical rotor with blades fixed to it. This is in an outer casing (unlike the Waddle fan). As the fan rotates, foul mine air enters the casing through one or two inlets. The rotation of the fan sets up a centrifugal force which spins the air outwards. The casing directs the air to a 'chimney' (or evasée) through which it wastes to the atmosphere. The vacuum continuously created in this way draws the air from the workings because the fan drift and upcast shaft top are made air tight to prevent air coming in from outside. In an emergency the fan can be used to force air into the mine.

There is another type of fan in use today in which the air flows parallel to the axis of the rotor. This is called the axial flow fan.

Sometimes, ventilation systems incorporated some unorthodox equipment. For example, at Harrycrofts Colliery, Kiveton, near Sheffield, which closed a few years ago, an aircraft propeller was used underground as an impeller fan. This was handed over to the Royal Air Force Museum, Henlow, Bedfordshire, in 1969. The propeller is in a good state of preservation, but so far it has not been possible to identify the type of aircraft from which it was taken.

By the end of the nineteenth century, mechanical fans had replaced furnaces at most collieries of any size; although strangely enough the writer of a book on coalmining published in 1896 still regarded the furnace as the most serviceable means of ventilation.[11]

The districts with the most pressing ventilation problems were those with fiery mines. Naturally, it was in these districts, principally Northumberland, Durham and South Wales, that the greatest progress was made with fans. South Wales was backward in this, as in most other respects, until mid-nineteenth century. However, as the gassy steam coal seams became more intensively worked, serious explosions so increased the hazards and costs of mining as to spur colliery owners on to improve their ventilation systems.

The first Struvé air pump was used at Eaglesbush Colliery, near Neath in 1849, and several similar installations followed. Similarly, the centrifugal fans invented by Brunton and Waddle were pioneered at Welsh mines. Even so, ventilation furnaces continued in general use in this coalfield until the last quarter of the century.

Improved ventilation reduced the risk of explosions, but naked flames, shotfiring with gunpowder, and coaldust were still major hazards.

Safety lamps came into general use at very few mines in the first half of the nineteenth century. Even in Northumberland and Durham, naked lights were still used in the less fiery mines. With the passage of time, they became more widely adopted; but it usually needed an explosion to convince owners and men that they were necessary.

The reluctance to use safety lamps was due partly to the comparatively poor light they gave. As illuminants they were not nearly so good as candles. At many collieries in the Midlands the changeover to safety lamps did not take place until the early twentieth century; the men then demanded, and were paid, an enhanced tonnage rate to compensate for the lower productivity caused by the poorer light. (Indeed a few shallow licensed mines still use candles.) The safety lamp was also responsible for the alarming increase in miners' nystagmus, a distressing and disabling complaint of which one symptom is oscillation of the eyeballs. Men working with naked lights hardly ever suffered from it, and since electric cap lamps have become almost universal the disease has completely disappeared.

Even in fiery mines, for many years safety lamps were used only in places or at times thought to be especially dangerous. Certain districts of a mine were known to be more susceptible to 'blowers' of gas than others; and the return airways were obviously more likely to be contaminated with gas than intake airways. Barometric pressure was also recognized as having an effect: a fall in the barometric pressure was concomitant with an increased emission of firedamp from the seam. This was observed as early as the late eighteenth century by Dr William Brownrigg, who drew firedamp from a neighbouring mine for experimental use at his Whitehaven laboratory. He noticed that supplies of gas increased when the barometer fell, and decreased when the barometer rose. By this means he was able to predict when explosions were likely to occur.

Another time of danger was on Monday morning at collieries where the furnaces were allowed to go out at the weekend. The following extract from the rules at Pendleton Colliery, Lancashire, in 1832 points to this danger:

> The overmen and deputies must thoroughly understand what is generally termed trying the candle, in parts of the mine where it is expected inflammable gas may accumulate, to ascertain whether it is near the firing point or not; if in any degree dangerous none must be allowed to go there without using the safety lamp and not the naked candle. The parts most likely to be charged with inflammable gas will be on the rise side of the workings, it being much lighter than the common atmosphere. It will require their greatest attention on the Monday morning, or when the pit has not been worked for a few days.[12]

Even as late as 1883 (and with mechanical ventilation) an explosion occurred at Brinsley Colliery, near Eastwood, in this way. The fan had been stopped over the weekend, allowing gas to accumulate. When the fan restarted, this gas was drawn towards the pit bottom. A deputy and an ostler went down the pit within an hour of the fan's restarting, and when the deputy lit a torch an explosion took place. This was so violent that it dislodged the headgear, blowing parts of it high in the air.

Trying with the candle was superseded by testing with the safety lamp. As with the candle, a blue 'gas cap' forms over the flame when firedamp is present; and if the wick is adjusted for testing purposes, the percentage of gas can be estimated very much more accurately than with a candle. In the middle of the century many mines kept one lamp for testing. This was the case in Derbyshire and Nottinghamshire in 1842, for example.

Safety lamps were mentioned in the 1872 and 1887 Coal Mines Acts, but in such a way as to leave their use voluntary. Thus the 1887 Act specified that no lights other than locked safety lamps were to be used where there was likely to be an inflammable mixture. This was so vague and indefinite as to be meaningless. The 1911 Act, on the other hand, specified far more clearly under what conditions the use of locked safety lamps was to be compulsory; namely, in any mine where there was normally over 0·5 per cent of firedamp in the main return airway, and in any seam where there had been an explosion causing personal injury during the preceding twelve months. The 1911 Act, therefore, divided mines into those where safety lamps were to be used and those where naked flames were still permissible. Candles were still in use at some quite large 'naked light' mines (Wollaton in Nottinghamshire, for example) as late as 1947.

While the candle was the most popular illuminant in the nineteenth century, there were other naked lights too. Crude tin lamps of local manufacture burning colza oil were popular in some places in the first half of the century. In 1847, however, a Manchester chemist, James Young, distilled petroleum found in a well at one of James Oakes and Company's coal mines at Riddings, Derbyshire, to produce paraffin. He patented his process in 1850 and from then Young's Paraffin Oil sold in great quantities. Miners found that it was a much more satisfactory illuminant than colza. Candles made from paraffin wax also competed with tallow candles. Wax candles, besides being cheaper than tallow, were better for sticking in a tin holder with a shield at the back. This method of using candles was novel in most places. The traditional method was to stick the candle in a piece of clay and attach it to a pit prop though there was always a danger that the prop might catch fire.

Old naked flame lamps are found much less frequently than safety lamps. Presumably this is because most of them were very crude and were thought to be not worth saving. One specimen which came to light recently in curious circumstances was, however, beautifully made. This is a 'Smokey' lamp with a wire gauze spout now owned by R. W. Storer. During repair work on the chimney of a house at Pinxton following mining subsidence, this lamp clattered into the hearth in a cloud of soot. Most 'Smokey' lamps were very much less well made. This is true, too, of most examples of the other popular type of late nineteenth-century lamp, the 'Midgie', which was still used at some pits within the memory of old retired miners.

In the early twentieth century, acetylene (carbide) lamps were used at many naked light mines, and they were still used at Shortwoods, a small mine in Shropshire until early in 1970. These gave a very intense light, but were not without their drawbacks. The acetylene reduces carbon dioxide to carbon monoxide, and this has been known to cause deaths where ventilation was poor.

Even where flame safety lamps were used, they could still be a potential source of danger. In his report for 1851 Charles Morton, one of the mines inspectors appointed under the terms of the Coal Mines Act of 1850, pointed out that:

> Davy recommended a wire gauze, containing not fewer than 784 apertures in a square inch; he advised that the framework and fittings of the lamp should be so arranged as to prevent the possibility of there being a larger external aperture in any part of it; he warned the miner not to expose it to a rapid current of inflammable air unless protected by a shield half encircling the gauze; and he deprecated the practice of continuing to work with it after the wire attained a red heat.
>
> In my visits to the collieries I have occasionally observed and pointed out to persons present their practical disregard of the forgoing conditions.
>
> Lamps are now made which contain only 550 or 600 apertures per square inch; the framework is in some instances so loosely put together, or so far dilapidated, as to exhibit even larger openings; shields are seldom met with; and the gauze is seen at times red hot, and smeared outside with grease and coal dust, whereby the risk of ignition is rendered more imminent.

He went on to say that in the Midlands safety lamps were provided only 'for experiment and trial', but in Yorkshire, where the seams were more fiery, they were used more frequently and some pits had no naked lights at all. However, despite the danger, the men were 'sometimes tempted to take off the gauze' in order to obtain a better light. He then discussed the relative merits of the various types of safety lamps available. The Clanny lamp (where the flame was surrounded by a glass cylinder) gave more light than a Davy; and so did the Biram lamp which had a reflector behind the wick. Both these lamps had the additional advantage that they went out when any considerable body of gas was present. The Upton and Roberts lamp, which was safer than any of the others, was little used because it was difficult to keep it burning.

After recommending that all safety lamps should be fitted with locks and should be examined by an official every day, he said: 'Boys need not be intrusted with safety lamps; fixed lights, stationed at intervals along the tramways, will be found sufficient for the performance of their duties; and blasting with gunpowder, the smoking of tobacco, and the use of lucifer matches, should be prohibited where safety lamps are employed.'[13] Gunpowder was by this time in general use and its beneficial effect on productivity was such that there was little hope of dispensing with it in fiery mines except by substituting a safer explosive.

From about 1870, explosives based on dynamite were developed for mining work although they were not widely used until the late 1880s. In 1886 a Royal Commission Report recommended the use of water cartridges to lessen the danger from shotfiring. This idea was not given practical form until the 1950s. Instead, 'flameless' gelatinous explosives were developed, and these have undergone a continuous process of change and improvement. In 1897 a government testing station was established. This had the duty of issuing lists of approved explosives which had passed tests as to safety.

Black powder continued to be used for a very long time, but even this changed its composition. Ingredients were added to the original gunpowder to make it safer to use and the resultant was called Bobbinite.

Electrical shotfiring equipment was developed in the 1860s, but 'safety' fuses of the old type, lit with a match, were widely used until 1910. Since that time nearly all shotfiring has been by electric battery except in small 'naked light' mines.

The dangerous nature of coaldust went unrecognized for a very long time. Indeed, when Dr William Galloway presented papers to the *Proceedings* of the Royal Society demonstrating that coal dust and air mixtures are potentially explosive, he was forced to resign his position as a mines inspector. Shortly after this, a Royal Commission report (published in 1893) accepted the fact that coaldust plays a major part in nearly all explosions and that an explosion can take place where there is no firedamp present at all.

Firedamp and air mixtures are, it is true, more readily inflammable than coal dust and air mixtures; but once a coal dust explosion has been triggered off, it gathers force from further accumulations of dust along the roadways and is much more destructive than a pure firedamp explosion.

Early attempts to prevent coal dust explosions bring us back to the subject of ventilation. Improved ventilation had had the effect of producing drier conditions underground. This is in one way a positive benefit to miners. Where there is a high temperature, as in most deep mines, it is important to keep down the humidity. Humid heat is much less tolerable than dry heat, particularly for men doing a hard physical job. However, improved ventilation, in producing drier conditions, had made coal dust more dangerous. At many collieries this problem was first tackled by spraying the roadways with water. But this method was expensive and it had physical drawbacks too.

A more practical solution, suggested by William (later Sir William) Garforth, was that stone dust should be mixed with the coaldust in the roadways. After various experiments carried out before and during the First World War it was made compulsory for stone dust to be applied on colliery roadways except where they were wet. More recently, stone dust barriers have come into use. It is therefore highly unlikely that there will be any more major explosions like those which occurred so often in the nineteenth century.

Of course, many of the victims of a mining disaster died not as a direct result of the explosion itself but by reason of poisoning by the third main gas met with in mining, 'after damp', the chief constituent of which is carbon monoxide. This results from combustion whether or not the combustion was associated with an explosion. Any underground fire can thus lead to a loss of life from carbon monoxide poisoning.

Recently it has become compulsory at many pits for all persons going underground to carry a 'self rescuer'. This is a device designed to enable its wearer to travel through 'after damp' (or other gases) to the pit bottom. As self rescuers become available, they will be issued to all miners.

So far there have been no reports of workings in which explosions have taken place being exposed during opencast operations. Most major explosions have occurred in deep mines which are unlikely ever to be opened up again. It is quite possible, however, that the sites of some minor explosions will be revealed in this way. Sometimes pit props showing signs of charring are found on opencast sites, but it has not been possible so far to identify these positively as being due to explosions.

Underground fires have been mentioned. Some seams are

particularly liable to spontaneous combustion. This is true of the coal mined at Brora Colliery, for example. The Colliery was, indeed, forced to close for a period by a recent underground fire. The Yorkshire Barnsley Bed is also notoriously liable to spontaneous combustion and so are some seams in Staffordshire, Shropshire and South Derbyshire.

In the first half of the nineteenth century the owners of such collieries were often unwilling to provide efficient ventilation because they feared the effect of a good supply of oxygen. The mines inspector whose report for 1851 we have quoted before had something to say on this:

> In sundry collieries of the Midland Counties, a spontaneous combustion of the refuse coal left underground occasionally breaks forth, which is said to be promoted by the free admission of the atmosphere, and thorough ventilation is, therefore, opposed by some of the viewers; but the plan so successfully practised by Mr. Woodhouse at the Moira Mines in Leicestershire, of excluding air from the 'goaves' by means of vertical clay puddling, demonstrates that complete ventilation can be safely accomplished, even where the much dreaded 'breeding fire' exists.[14]

Where spontaneous combustion does occur, it is dealt with either by excluding air from the seat of the fire; or by digging out the burning material.

Just as some seams are particularly liable to spontaneous combustion so are some (sometimes the same ones) particularly noted for the emission of methane. In some cases far more methane is given off than could be effectively cleared by ventilation alone, and methane drainage systems are installed.

Methane drainage was pioneered by James Ryan in the first decade of the eighteenth century. He drove levels in the roof of the coal seam and connected these with the upcast shaft. This system of drainage was successful in the comparatively small mines working the South Staffordshire Thick Coal, but was thought to be useless for the extensive collieries of Northumberland and Durham.

As long ago as 1839, methane was brought up the 'C' Pit at Wallsend Colliery through a four inch diameter metal pipe. According to a contemporary observer, this gas came from a particular area of coal of about five acres and was

> carried, by means of the pipe above mentioned, to the top of the head gear, where it burns with the utmost fury. This dreadful oriflamme waves its fierce streamers in the wind to the extent of eight or nine feet of flame; and it sounds like the roaring of a blast furnace. It has been computed that the quantity of gas drawn off from the mine, and consumed by this means, is *eleven hogsheads per minute*! amounting to the enormous quantity of 15,840 hogsheads per day ![15]

This is illustrated by T. H. Hair in his remarkable *Sketches*. A foot-note reminds us that the same method of methane drainage was adopted at Workington and Whitehaven many years earlier and that the colliery agent there had offered to supply Whitehaven with gas from the pits.

In modern mining practice, methane drainage is practised on a large scale. The gas is sometimes used in the colliery boilers and sometimes piped to a nearby works of some kind. Thus, gas from Harworth Colliery in North Nottinghamshire is sold to the nearby factory of Glass Bulbs Ltd.

CHAPTER SIX

Drainage

In any technical history, the subject matter has to be divided in some way for the purpose of exposition. Galloway, in his *Annals of Coal-mining and the Coal Trade*, the first volume of which was published in 1898, adopted a chronological division. Within this broad division into periods, he divided his history further by districts. Here we have adopted a functional division instead. However, it must be understood that any division is artificial. Thus, winding, ventilation and drainage, for example, go on simultaneously. In this chapter, we shall have to go over some old ground because these three things are interconnected. Some of the equipment used for one purpose is used also for another. This is true of early winding gear, which was used for drainage as well as winding coal out of the pit. Less obviously, pumping water up a downcast shaft helped natural ventilation by keeping the shaft (and the air in it) cool. Conversely, where pumping took place in an upcast shaft, as it did at some mines in the 'furnace' period, ventilation was retarded.

Water was, for centuries, the miner's biggest problem. The extent of the problem for medieval miners working on the outcrop can be gauged by walking through a ploughed clayey field after heavy rain or after winter's snows have thawed. Even at the end of August 1969, when there had been little rain for some time, some of the bell pits exposed at Stretton had several inches of water standing in them. The same site had been completely waterlogged six months earlier; and indeed the contractor had been forced to suspend operations until conditions improved.

In the earliest period of mining coal from the outcrop, drainage was achieved by ditching. In the nature of things, ditches normally leave no trace for the industrial archaeologist but on at least one opencast site a line of bell pits interconnected by a trench has been found. It seems likely that the ditch was first dug through the field and that the pits were then sunk along its length.

Where the hand windlass was in use for raising coal it could also be used to raise water in buckets. This is no doubt why it is usually called a wallower (corrupted to 'waller') in the East Midlands. In Derbyshire there were still some small mines being drained by 'two men at a waller' in the 1840s. A ruined wallower was seen by Mr Frank Nixon at the site of the eighteenth-century Yatestoop Lead Mine, near Winster, Derbyshire, a few years ago, where it no doubt remains to protect the owner's rights in the mine. Until the invention of the steam engine, improvements in drainage techniques were essentially improvements in these two things; ditching and winding water in buckets.

Thus, as mines became deeper 'ditches' had to be driven underground, and buckets became attached to machinery. The underground ditches were called 'water gates' in the north-east; 'adits' in southern England; and 'soughs' (variously spelt and variously pronounced) in the Midlands. A lease of coal at Gateshead dated 1364 mentions a water gate for draining the workings; and so does another Durham lease dated 1354. If these were driven underground they were the earliest ones of which we have positive evidence.[1] By the middle of the sixteenth century, some long soughs were being driven like the one over a mile in length, which is said to have cost the Willoughbys of Wollaton £20,000 in 1552. An even longer east Midlands example is a sough draining mines in the Blackwell and Teversal areas (on the Notts–Derbyshire border) which was driven in stages between 1703 and 1774 reaching a total length of over five miles. This last was a joint drainage scheme to which several colliery owners contributed. There is still some archaeological evidence of this sough. A northern example, dating from the seventeenth century, was the Delavel Drift which drained workings on the Benwell Estate, Whorlton, Newbiggin and West Kenton.[2] There were many others in the Great Northern Coalfield. And for every long sough, there were dozens of short ones.

All early soughs, and most later ones, were driven in the coal seam itself. All the workings on the rise side of the sough were thus drained by it. Generally, when a new pit was sunk, a water level was driven from it to the nearest sough. A great deal of unpleasantness, and some costly litigation, was caused by the fact that many mineowners drained their water into soughs driven at considerable expense by others. In the Forest of Dean it was illegal for one miner to sink a pit within 100 yards (extended to 300 yards in 1692) of a sough driven by someone else unless he had obtained permission. An example of the sorts of disputes over soughs which arose in all coalfields is illustrated in Nef's book *The Rise of the British Coal Industry*. This dispute affected several coal owners in the Erewash Valley who alleged that one of their competitors (John Fletcher and Partners, later styled Barber, Walker and Co.) was stopping up soughs so as to drown their workings, thus giving him a monopoly of the coal supply in the district. John Fletcher denied this strenuously. However, the other owners introduced a parliamentary Bill to make it illegal to destroy coal works. The dispute may be best understood from Fletcher's reply, which also gives valuable insight into eighteenth-century mining practice:

Case of the Petitioner John Fletcher, and his Co-partners

AGAINST

The Bill for further and more effectually preventing the wilful and malicious Destruction of Collieries and Coal-Works.

The Reasons given for the bringing in of this Bill are; That the Petitioner and his Partners have several Collieries in Derbyshire and Nottinghamshire as appears by the Plan; and they drove, or headed in their Colliery at Smalley, in Derbyshire, into the Old-Level to let Wind into the Work to prevent the Damp; which is a usual and necessary Practice in the Working of Collieries. They also sunk Pits upon their own Land to get their own Coal, and did get, and might have got, great Quantities of Coal, out of them. They also stopped up one old Sough, upon their own Ground; which could not have drowned Mr Richardson's Work at Smalley, had his Fire-Engine been in Repair, the Water not being above twenty yards deep in the Engine Pit, when the Engine was set down; which if it had been in any tollerable Repair would have drawn all the Water. However, all that the Partners did was upon their Ground, and for the better carrying on their own Work.

Mr Richardson, indeed, did bring an Action against the Partners for letting in the Water, and recovered a Verdict for 200*l*. Damages: which Verdict could never have been obtained, had not the principal witnesses on the Partners behalf been designedly made Defendants. However, this Verdict shows Mr Richardson has a legal Remedy for any Damages he may sustain.

THE WATER whereby Mr Richardson pretends he received this Damage is not a continued Water-Course, but is only what is called a Land-Flood, and only runs when it Rains; and there is no Colliery whatsoever but the Ground will break and let in the Land-Flood, in some measure, after great Rains; To prevent which, as much as possible, the Petitioner and his Partners, both before and since this Verdict have made Ditches upon fresh Ground to carry off all the Water, so that except Mr Richardson has himself let in the Water on purpose to lay a Foundation for this Bill, it is impossible he can be drowned by the Land-Flood.

Another Complaint made against the Petitioner, and his Partners, is from Mr Lowe, who is possessed of Denby Colliery; And the Fact is, That the Partners rented that colliery of his Father, and had a Lease for his Life; and during the Time they were possessed of this Colliery, they drove a Level from their own, called Roby's Colliery (which was before unwatered by Loscoe Sough) to Mr Lowe's colliery, which unwatered it. After Mr Lowe the Father's Death, the Petitioner, and his Partners, stopped up this Sough thus made by themselves, at their own Expence, upon their own Ground; by which they left Mr Lowe in no worse but a better Condition, with respect to Water, than he would have been in case that Sough had never been made.

As to the Enhancing the Price of Coals: It will be time enough to make an Application to Parliament when the Petitioner, and his Partners, are guilty of that Offence, which is so far from the Case at present, that notwithstanding the Expence of getting Coal in Wages, and otherwise is much larger than ever, yet Coals now are as cheap as they have been for twenty years last past to the Consumers. And by the Number of Collieries in the Plan, it may appear, that the Petitioner, and his Partners, have not a fourth Part of the Collieries in Derby and Nottinghamshire, worked and capable of being worked, and therefore there is no Danger of a Monopoly.

That the Petitioner, and his Partners, and those under whom they claim, have been at near 20,000*l.* Expence, in making Soughs and other expensive Work to drain and work their Collieries; and now are under Leases, and stand liable to pay very large Rents to Persons thro' whose Ground those Soughs are cut, and for their Coal-Mines. And if they are not to be at liberty to stop up these Soughs upon their own Ground, then the Consequence will be, That other Persons who have contributed to no part of the Expence will reap all the Benefit, by having their Collieries laid dry at the Expence of the Partners; who will not be able to pay their Rents. And if other Persons are to have the Benefit of the Soughs without having contributed to any part of the Expence, or being liable to pay any part of the Rents, there will be a greater Danger of Monopoly in them than in the Partners.

If this Bill passes, it will be a total Discouragement to driving of Soughs; for who will be at such an Expence, when upon the Expiration, or some other Determination of a Lease, he can have no Power to stop them up, but some other Person must have the Benefit of them, and if no body will drive soughs then Coals must be got by Fire-Engines; which being a much more expensive way of draining, will considerably increase the Price of Coals, and leave Mines fuller of Water, and in a more ruinous Condition than they can be by stopping of Soughs.

Lord Middleton's pits, near Nottingham, are contiguous to the Petitioners Pits at Kimberly and Bilburrow; and his Complaint is, That the Petitioners have attempted to drown him, by stopping their Sough in Kimberly.

As to this; The Petitioner, and his Partners, are also possessed of a Colliery adjoining to Kimberly, belonging to Lord Stamford, under a large Annual Rent, and the Sough, which Lord Middleton complains they have attempted to stop up, runs from their Colliery at Kimberly into that which they hold of Lord Stamford, which fills that Colliery full of Water. The Petitioner, and his Partners, therefore apprehend they have a Right to stop up Kimberly Sough, merely to enable them to work Lord Stamford's Colliery.[3]

Workings in one seam were sometimes connected to a sough in a lower seam by means of a staple shaft. Similarly, water from a seam below the level of a sough was sometimes wound up a staple in buckets (or, later, lifted by pumps) to the sough. Some soughs took the form of cross-measure drifts, that is, instead of being driven more-or-less horizontally in one seam, they were driven from the level of one seam obliquely down to the level of another.

The mouths of many old soughs can still be seen, one of the most famous being the Haigh sough which drains into the Yellow Brook north of Wigan. It is often difficult to distinguish sough outfalls from culverts; but in Shropshire they can usually be identified by the ochrous water which flows from them. Soughs are also exposed from time to time by opencast operations. One such, probably mid-eighteenth-century, was revealed at Newman Spinney near Barlborough in 1965/6. This was driven in the Clowne Seam at a depth of approximately 180 feet. All the coal worked lay between this sough and the outcrop of the seam, approximately 300 yards to the West. On the dip side of the sough the coal was untouched. This sough was supported by a single row of props which appeared to have been set (as props usually were in early mining practice) whilst in the 'green' state. Some soughs were supported with props and bars and some (mainly nineteenth century ones) were bricked. Most Shropshire soughs appear to have been unsupported.

How were soughs driven? According to Greenwell,[4] some North country soughs were no more than eighteen inches wide yet their sides were as smooth as if they had been chiselled. It is difficult to see how a man with a pick could drive such a narrow level; and it is inconceivable that a boy would take such pains to smooth the sides, In the absence of any confirmatory evidence, one finds it difficult to accept Greenwell's story. Ivor Brown suggests that these soughs were eighteen inches wide at the bottom but wider at the top to accommodate the man's shoulders, as with some Derbyshire lead mining soughs. Soughs were certainly narrow; and where they were driven in thin seams they were low also. The only way in which many of them could have been driven was by a man lying on his side and wielding a pick. The debris was no doubt pulled by a second person—probably a boy—to the end of the drift. Ventilation was provided, as we have seen, by forcing air through wooden pipes by a bellows or (at a later date) by a hand-driven fan. With the very long soughs like the one in the Blackwell/Teversal area, shafts were sunk at intervals to facilitate a flow of air.

Generally soughs were only of use where the workings were above the level of free drainage. Sometimes, however, short water-levels were driven not to an outfall but to a sump pit. A sump pit was one sunk to a greater depth than the others. Water drained from the workings into the sump and from there was raised to the surface by a device of some kind. Even with a good sough, the flood water of the winter months might stop production. Sometimes, indeed, winter flooding forced water back up the sough into the workings.

Soughs were expensive to drive. Many small mineowners could not afford to make them and instead they wound the water out in buckets or tubs. Even where there was a sough it was sometimes necessary to supplement this by winding water. And the really deep mines of the north-east, being below the level of free drainage, had no option but to wind water up a shaft.

J.C. in his treatise on mining, *The Compleat Collier*, written in 1708, explains that, in sinking a shaft it was sometimes necessary simultaneously to sink a half-shaft (i.e. a narrow shaft) adjacent to it, keeping the half-shaft always deeper than the main shaft so that any feeders of water encountered in sinking could be run off into the half-shaft and then wound to the surface. Water could also be 'framed back' during sinking by using fir tubbing with sheepskins, wool side outwards, between the tubbing and the stone work.

As to the method of winding water, J.C. explains:

> we generally Draw it by Tubs or Buckets; whilst Sinking with Jack Rowl, or by Mens winding up the Rowl, or otherwise, if the Pit be Sunk more than thirty Fathom, then we use the Horse Engin, which Engine being wrought with one or two Horses at a time, as the Water requires, serves also, after we have Coaled the Pit, to draw up the Wrought Coals. . . . In some places we draw Water by Water, with Water-Wheels or long Axel-Trees, but there is not that Conveniency of Water every where, and as for Mind-Mills, or Ginns to go by Wind; 'tis sure the Wind blows not to purpose at all times.

Drawing water by tubs or large buckets continued until well into the nineteenth century at many places, and there have been some examples in this century. A clay pit at Blists Hill, Shropshire, still used this system in 1950. At Brora, as we have seen, the water was wound in a tub to the top of the headframe where an automatic trip tipped it up, its contents pouring into a large tank from which it was piped to the sea. Evidence of this fairly recent practice is to be found in the truncated headstocks of the upcast shaft to which we have referred. These were demolished in late 1969. Similar tubs were used in the north-east in the mid-nineteenth century and were described in a mining engineering text book of 1848. Again, when the Molyneaux Colliery at Teversal, Nottinghamshire, was flooded out in 1869 a large tub, operated by a borrowed steam engine, was used to clear the shaft. A late nineteenth century water tub from Pinxton, now at Lound Hall, is an ordinary wooden coal tram with a bung in its base.

Clearly tubs were simple and useful. Otherwise, they would surely have been replaced by one of the alternative horse-driven devices some of which were available from a very early date. One of these was itself no more than a chain of buckets. Here, an endless iron chain with many wooden or leather buckets attached to it was suspended in the shaft over horizontal axles at top and bottom. This was constantly revolved, the links of the chain meshing with a cog on the axle. Water was thus scooped up in the bottom of the shaft (the 'sump') and discharged into a channel on reaching the surface. This device was in use at some places (mainly in Scotland and the north of England) in the sixteenth century.

Similar to the chain of buckets was the rag and chain pump popular in the Midlands. Here, instead of buckets the chain had a series of discs, probably made of brass, affixed to it. These were drawn through a wooden cylinder which extended from the surface into the sump. As the chain revolved, water was carried up the cylinder by the discs.

Another device patented in 1619 also used a wooden cylinder in the shaft, but here a rod with a leather-covered piston at the end took the place of the chain. As the piston was raised in the cylinder, water rushed into the cylinder from the sump to fill the partial vacuum thus created. The piston was designed to allow the water to flow past it on the downstroke. On the second upstroke, the water above the piston was forced up the cylinder and a fresh inrush from the sump occurred below the piston. Once the pump was primed, a continuous stream of water issued from the top.[5]

Huntingdon Beaumont used horse engines (almost certainly rag and chain pumps) in the first decade of the seventeenth century; and when he left Wollaton temporarily for the north, he introduced his engines there. If Gray is to be believed, Beaumont's engines were new to the Northumberland–Durham coalfield and were regarded as an improvement on their existing devices. But they were certainly not new in Nottinghamshire. One was in use at Wollaton in the mid-sixteenth century.

Both waterwheels and windmills were used to drive some of the Wollaton pumps; but, as J.C. observed a hundred years later, they were not to be relied on. Writing to Sir Percival Willoughby, Beaumont reported on one occasion that 'the water wheele doth so well play her parte as that I thinke it can not be bettered'. But even then, there were fifty-six pump horses at Wollaton. Of these, twenty-six were out at grass, no doubt because the waterwheel was, for the moment, working so well.[6]

Whether worked by waterwheel, windmill or horses, these early pumps could lift water for only about thirty or forty yards. To drain deeper pits, the water had to be lifted in stages from one level to another.

There is no doubt that the lack of efficient drainage devices limited the depths to which pits could be sunk. That is one reason why thin, poor seams near the surface were sometimes worked whilst better seams lay untouched at lower levels. Flooding also caused many small mines to be abandoned, and made work impossible at others in the depth of winter.

The first attempt to drain mines by steam power, Thomas Savery's engine, was a failure. This engine was invented in 1698. Essentially, it consisted of a boiler connected to a cylinder by valves. Steam generated in the boiler was fed into the cylinder and was then condensed to create a vacuum. Water from the sump rushed into the cylinder to fill the vacuum and it was forced up the rising main by live steam. This engine would only draw water from depths of sixty or eighty feet, besides which it had no form of safety valve. Any strain may, therefore, have caused the boiler to explode.

J.C. had heard of it, but treated it with scepticism:

> If it would be made Apparent, that as we have it noised Abroad, there is this and that Invention found out to draw out all great old Waists, or Drowned Collieries, of what depth soever; I dare assure such Artists, may have such Encouragement as would keep them their Coach and Six for we cannot do it by our Engines, and there are several good Collieries which lye unwrought and drowned for want of such Noble Engines or Methods as are talk'd of or pretended to, yet there is one Invention of drawing Water by

> Fire, which we hear of, and perhaps doth to purpose in many Places and Circumstances, but in these Collieries here a way, I am affraid, there are not many dare Venture of it, because *Nature* doth generally afford us too much Sulpherous Matter, to bring more Fire within these our deep Bowels of the Earth.[7]

Savery's engine had to be housed in an inset in the shaft within twenty-six to twenty-eight feet of the sump. In his treatise, *The Miner's Friend*, Savery suggests that the furnace of the engine would serve to provide ventilation in addition to its primary purpose. He says:

> The far greater part of our richest mines and coal-pits, are liable to two grand inconveniences, and thereby rendered useless, viz., the eruption and excess of subterraneous waters, as not being worth the expense of draining them by the great charge of horses, or hand labour. Or, secondly, fatal damps, by which many are struck blind lame or dead, in these subterraneous cavities, if the mine is wanting of a due circulation of air. Now, both these inconveniences are naturally remedied by the work of this engine, of raising water by the impellent force of fire.[8]

Whilst this device was unsuccessful, the Newcomen engine invented shortly afterwards justified fully Savery's sanguine belief in the efficiency of 'the impellent force of fire'. Thomas Newcomen was a blacksmith or ironmonger by trade. He applied his practical knowledge and skill to the theoretical concepts of others (e.g. Papin) to produce the first successful engine for draining mines. Crude as his early atmospheric engines were, they represented a magnificent achievement. And they worked.

Newcomen's engine (often referred to by contemporaries as the 'Fire Engine') was brought out under Savery's patent of 1698, although in two important respects it was quite different. First, it has a piston, and second, it was fitted with a safety valve and other gear. In early models a brass cylinder housed a brass piston surmounted by a soft leather flap kept soft and airtight by a layer of water which always lay on top of it. Later models had cast iron pistons with hemp packing kept soft, as with the earlier leather cap, by a layer of water. The boilers were originally made of copper, but later ones were made of wrought iron. The movement of the piston in the cylinder was communicated to a massive wooden beam, pivoted in the middle, which moved up and down like a pump handle. Indeed, the beam was a pump handle, simulating the actions of manually operated ones. Water was drawn from the sump through a continuous pipe running from top to bottom of the shaft by the action of pump rods moving up and down.

Steam generated in the hemispherical boiler was admitted to the cylinder, beneath the piston, at quite a low pressure. This steam being condensed by a jet of cold water, a partial vacuum was created beneath the piston. The top of the cylinder was open so that atmospheric pressure forced the piston down to fill the vacuum. This downward motion of the piston was transmitted via the wooden beam (the pump handle) to the pump rods in the pit shaft whose weight then lifted the piston for the next stroke. The process was repeated in a continuous cycle. It was a beautifully simple action, but it is not true that early engines could be left entirely in the care of young boys. They required three men to start them. Further, because of defective materials and workmanship, they required the services of a skilled engineman to keep them running. Later engines, made with greater precision, were much easier to run.

The first Newcomen engine to go into service at a coalmine was erected in 1712, almost certainly at Coneygree Colliery, Tipton. This had automatic valve gear.[9]

It must be appreciated that in the early eighteenth century, there was no engineering industry as we know it today. Particular engines made use of the skills of local craftsmen, blacksmiths, plumbers, millwrights, and so on. The cylinders were never true, although when the Coalbrookdale Company began to undertake this work their accuracy improved. Modifications to the original design were made by Newcomen (who died in 1729) and others; and standards of materials and workmanship similarly developed. By the time the Farme Colliery engines mentioned earlier were built, engineering standards were very much higher than in Newcomen's day.

The pumping engine erected at Farme Colliery in 1821 had a sixty inch cylinder and a stroke of seven feet. The total lift of the pumps was 140 yards in three stages. Steam was supplied by two haystack boilers. The piston was a crude affair, packed with old hemp rope (previously used in the pit) to make a reasonably tight fit. As noted in Chapter Three, 'Shafts and Winding', this engine was dismantled in 1888.

Some other interesting Newcomen engines used at collieries near Stourbridge are illustrated in the *Proceedings of the Institution of Mechanical Engineers* for 1903. One, the Buffery Old Colliery pumping engine, had a separate condenser on the Watt principle; most atmospheric engines were modified in this way following Watt's invention.

Another engine illustrated in the *Proceedings* was still *in situ*, though in a ruined state, in 1903. This was at Bardsley, near Ashton-under-Lyne, where it had been used as a pumping engine at the 'Camel' pit. This was a very early engine having a wooden beam with metal bracings. The cylinder was twenty-eight inch diameter, with a six foot stroke. The engine last worked about 1830. Fortunately a photograph was taken whilst it was still in a reasonably complete state, and this was deposited at the South Kensington Science Museum.[10]

Another pumping engine, built at Carron Iron Works about 1770 to 1780, was in use at Caprington Colliery from 1806 to July 1901. It pumped water from a depth of 165 feet. The cylinder was marked thus:

No. 31935
1–6–3–0
CARRON

This engine was taken over by the Corporation of Kilmarnock for re-erection in the Dick Museum.

Similarly an atmospheric engine of 1791 in use for many years at Pentrich Colliery, Derbyshire, was acquired by the South Kensington Science Museum in 1917. This was built by Francis Thompson of Ashover for Oakerthorpe Colliery, but was moved to the nearby Pentrich Colliery in 1841. Another atmospheric engine thought to have been built by Francis Thompson was installed at the Bassett Pit, Denby, Derbyshire, in 1817, after having been in use previously at another coal pit. It remained in use until 1886 and was photographed while still working. This engine was fitted with a pickle pot condenser. It had a piston of twenty-six inches diameter and thirty-nine inch stroke. Like all atmospheric engines it used steam at a low pressure: two pounds per square inch, and it operated at twenty-two strokes a minute. The photograph (which also appeared in the *Proceedings of the Institution of Mechanical Engineers*, 1903) is interesting because it shows two engines coupled together, one of them apparently being used as a whimsey. The haystack boiler shown on the photograph is almost certainly the one found a few years ago by Frank Nixon in a hedge at Park Hall Farm a mile or less away from the site of the pit. This boiler has since been taken to the South Kensington Science Museum.

Another old Newcomen engine, which worked a pump for 128 years, from 1795 to 1923, is still intact at Elsecar. The following description was supplied by the National Coal Board, South Yorkshire Area, whose Chief Engineer, Mr S. Collier, has been most helpful:

> The first beam was of wood with chains to couple the piston and the pump rods. Later, parallel motion was substituted and a cast-iron beam (supposed to have been cast at Elsecar Foundry) was put in. This beam is cast in two sections and is 24 feet in length and 4 feet 4 ins high in centre. The ends of the beam are fitted with stop pins.
>
> The early type of boilers supplying steam to this engine were of the 'haycock' or 'haystack' type. Later, two Cornish boilers, 22 feet by 7 feet were installed. There are no tubes or flues to these boilers and they are external fired. The working pressure is $1\frac{1}{2}$ to $2\frac{1}{2}$ lbs per square inch and the coal consumption about 18 to 20 lbs per H.P.
>
> The engine cylinder is 4 ft diameter with a 5 ft stroke and is water sealed to a depth of 9 to 12 inches.[11] The casting is about one inch thick.
>
> An unusual feature is a flat false bottom in the cylinder. This consists of a circular plate having a rectangular opening 11 in by 6 in just above the steam inlet.
>
> A $\frac{7}{8}$ inch diameter hole for the injection water is in the centre. The plate rests on a slight projection of the cylinder bottom flange and two long bolts to the sloping cylinder bottom hold it in position.
>
> The piston is packed with plaited spun yarn and is held in position by a junk ring.
>
> The engine makes six strokes per minute and with a total single lift of about 40 yards delivers 50 gallons per stroke, so that at its best the pump delivers 300 gallons per minute.
>
> The bucket with leather rings is 18 inches diameter.

Between the piston and the centre bearing are two rods (called Plug Trees) one of which works the 9 inch force pump which carries water (water pumped from the pit is used) into the injection supply tank, and the other rod actuates the 4 inch boiler feed pump. The indicated horsepower is 13.16.

The Newcomen engine was welcomed by the owners of large collieries whose workings had become waterlogged. An interesting example is Edmonstone Colliery in Midlothian. By 1725 this mine had become unworkable because of accumulations of water. Accordingly, the proprietors decided to install a Newcomen engine. The cost was surprisingly high. First, they had to pay a royalty of £80 a year for the licence to erect 'one engine, with a steam cylinder nine feet long and twenty-eight inches diameter'. The engine and associated equipment cost £1,007 11*s* 4*d* in materials alone, to which must be added the cost of scarce specialist labour and the cost of erecting an engine house. According to Robert Bold, who examined all the papers regarding this, the cylinder, buckets and clacks and some of the working barrels of this engine were made of brass, the boiler top was made of lead whilst the 'common pumps for the pit' were made of elm cut from the solid tree and banded with iron and having a bore of nine inches.

The value and scarcity of skilled men at this early stage in the development of an engineering trade, and the amount of attention required by the engines of the time, may be gauged from the payment made to the engineers responsible for the Edmonstone installation, John and Abraham Potter of Chester-le-Street. They were to be paid a minimum of £200 a year. In addition, they were to have half the clear profits of the colliery remaining after all expenses had been met. Further, had the engine proved unable to clear the mine of water, they were to be allowed to take away all the materials they had supplied and were to be recompensed for 'their pains and charges'.[12]

By the early nineteenth century, despite the price inflation of the French wars, a large pumping engine cost only about £2,000, and a whimsey (a much smaller engine) about £500. A Newcomen pumping engine installed at a new pit at Coleorton, Leicestershire, in 1784 had cost only £1,317, considerably less than the Edmonstone engine of 1725.

After Newcomen's death, the design of the engine remained substantially unchanged for forty years or more; but the materials used, and the standards of workmanship in construction, erection and maintenance underwent a process of continuous improvement. For example, by mid-century iron had replaced the brass, copper, lead and wood used in the early days. In the third quarter of the century the engines were widely adopted for pumping at collieries with drainage problems. They were by this time being made very much larger than the early engines; and they were not only less expensive to buy but, more important, much less expensive to maintain. An engine installed at one north country colliery (Walker) in 1763 had a bore of seventy-four inches, weighed over ten tons and required four large boilers (of which three were in use and the other on stand-by at any one time) to supply it with steam. It raised the water 178 yards in three lifts up to an adit which delivered it into the Tyne. The piston had a stroke of six feet, and operated at eight or ten strokes a minute; the boilers consumed about 6½ tons of coal a day.

Another engine, erected at Peggs Green, Leicestershire, in 1805, discharged 750 gallons of water per minute according to a contemporary observer. Incidentally, a nearby public house (now closed, but still occupied) was given the name 'the New Engine'. Names such as this are valuable clues for the industrial archaeologist.

John Smeaton, who had been instrumental in designing greatly improved water wheels for colliery and other work, turned his attention to the Newcomen engine in 1769. The original engine had been made by a practical man with no great grasp of the theory of his subject, and similarly the improvements made in the engine had been made on the basis of trial and error. Smeaton, by contrast, studied the existing atmospheric engine so as to derive general principles on which future engines should be designed and built. In this way he was able to improve the efficiency of the engine. One of Smeaton's engines was built in 1780 at Middleton Colliery, Leeds, which was the home of many improvements in technology. This engine had a cylinder seventy-two inches in diameter, and worked at nine strokes a minute with a nine foot stroke. Middleton Colliery with its headstocks, steam winding engine, and locomotive drawing a train of wagons, was depicted by George Walker whose paintings were reproduced in the *Costume of Yorkshire*, 1814.

The economic effect of improved efficiency was that the same quantity of water could be lifted with a much smaller consumption of coal. The practical effect was that one engine could cope with greater quantities of water. Despite Smeaton's improvements, however, the atmospheric engine was still wasteful of steam. How wasteful was discovered by James Watt in the winter of 1763–4, when he was engaged on repairing a model of a Newcomen engine belonging to Glasgow University. He found that the volume of steam supplied by the boiler for each stroke of the engine was sufficient to fill the cylinder several times over. The bulk of the steam was lost by condensation against the walls of the cylinder.

In 1765 Watt conceived the idea of condensing the steam in a separate vessel connected to the cylinder by a valve. The cylinder itself would be kept at the same temperature as the steam entering it. To achieve this, he proposed to employ a cylinder with a top. Whereas in Newcomen's engine, the atmosphere pressed on the top of the piston (the cylinder having an open top) in Watt's design, steam would take its place.

A detailed technical exposition of the Watt engine would be out of place here, but the following may help to explain the difference between Watt's beam engine and that of Newcomen. On the upstroke of the piston, steam was admitted into the bottom part of the cylinder from the boiler whilst the valve leading to the condenser remained closed. The steam pressure above and below the piston would then be about equal, but the piston would rise owing to the weight of the pump rods acting through the beam as with the Newcomen engine. Next, the valve connecting the bottom part of the cylinder to the condenser was opened while the valve controlling the entry of steam was closed. Steam therefore flowed into the condenser, thus producing a vacuum in the bottom part of the cylinder. The piston was therefore forced down to fill the vacuum by the weight of the steam above it. This movement of the piston was transmitted to the pump rods by the beam.

Watt patented his engine in 1768. He found practical difficulties in bringing it to a reasonable state of efficiency, however. His first partner, Dr Roebuck, was unable to supply as much capital as Watt had hoped and soon went bankrupt; and in addition the standards of workmanship required for the new engine were higher than for the simpler atmospheric engine and were difficult to meet. This is one reason why the Newcomen engine retained its popularity. Another reason is that the extra coal consumed by the Newcomen engine, as against the Watt, cost colliery proprietors very little. They therefore had no incentive to pay the high premiums demanded by Watt and his new partner, Matthew Boulton of Soho, Birmingham.

Watt was fortunate in his new partner. He was now able to draw on the skilled labour of the Birmingham district, and Boulton prevailed on John Wilkinson, ironmaster of Broseley, Shropshire, to produce the more accurate cylinders which the new engine required.

Watt improved his engine still further in 1782 when he patented the double-action principle. Here steam was applied not merely above, but also below the piston, with the vacuum alternating conversely. It was this improvement which gave the Watt engine its superiority for producing circular motion. Nevertheless, as we saw in Chapter Three, plenty of Newcomen engines were used for winding coal in the later eighteenth and early nineteenth centuries.[13]

The Cornish engine was developed from Watt's engine after the expiry of his patent. It is a single cylinder engine which retains the steam jacket of the early Watt engines. It employs steam at much higher pressures, working expansively; and is much more efficient and economical of fuel than earlier steam engines. Despite its superiority its adoption by colliery owners was surprisingly slow. The cost of the substantial foundations which it required may have been responsible for this. True, the engine was installed in a fair number of deep mines with particularly heavy drainage problems in the first half of the nineteenth century. Thus, 400 h.p. Cornish engines were installed at Pendleton Colliery, Lancashire, in 1843, and another large engine was purchased by George Stephenson for his new Clay Cross Colliery in 1846. But Galloway, writing in 1881, could still say that 'notwithstanding the great economy of the Cornish pumping engine, it has been comparatively little employed in the colliery districts'. By that time, however, these engines were being installed at coal mines in considerable numbers. When the mines were nationalized in 1947 a fair number were still in use; but most of them have since been scrapped.[14]

One drawback of the Cornish pumping engine was that it was not adequately protected from shocks caused, for example, by breakages of pump rods; although it was safer than the earlier beam engines.[15] In 1880 H. Davey largely overcame this defect when he invented

17 Braddyll Loco, built by Timothy Hackworth *c.* 1837. Note the fishbellied rails

18 Hard coal face, Clifton Colliery, 1895

19 Clifton Colliery again: the use of pick and ringer to break coal

his differential valve gear which automatically balanced the steam pressures so as to protect the engine from damage if a pump rod broke or if some other exceptional strain were thrown upon it. (Fig. 14.)

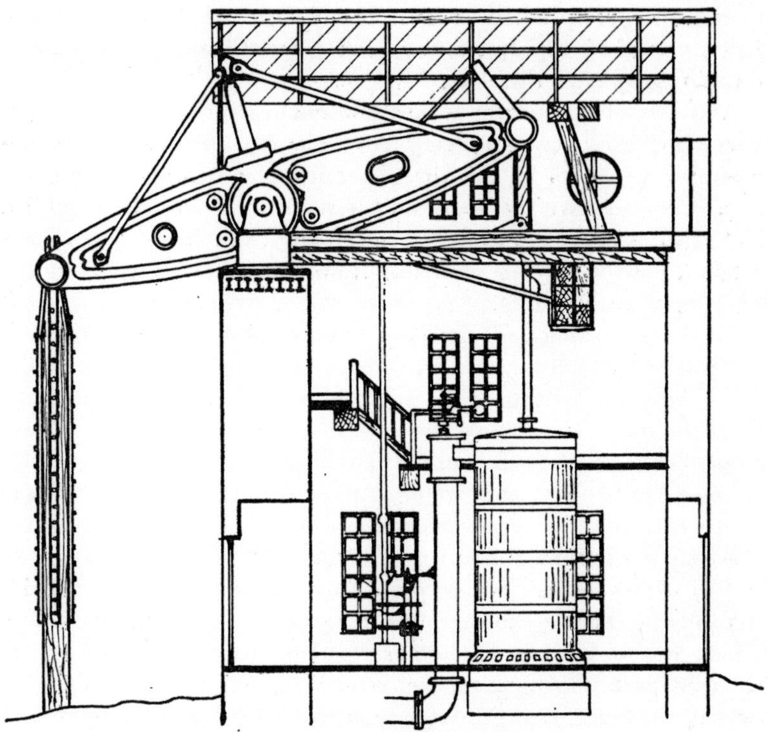

Fig. 14 Prestongrange beam engine

There are several Cornish Beam Engines still in existence, though in varying stages of dilapidation. One which is in a good state of preservation at a closed colliery at Preston Grange in the South Scottish Area is being taken over by the local authority who intend to make it into a central point of a mining museum. This is an excellent idea, and it is to be hoped that financial difficulties are not allowed to interfere with the plan. The beam of another Cornish engine is still *in situ* at Devon Colliery, Clackmannan.

In the late nineteenth century direct acting pumps came into favour. These occupy less space than beam engines and reduce maintenance costs because they do away with pump rods in the shaft. By this time, however, electricity was being introduced at many collieries and electrically driven pumps were preferred for siting underground rather than direct acting steam pumps. The first electric pump, installed at Trafalgar Colliery, Forest of Dean in 1881, was of 1½ h.p. but much more powerful models were introduced shortly after this date.

With direct acting force pumps powered by steam it was frequently necessary, in the case of very deep pits, to have several pumps mounted in insets in the shaft, because the steam pressure was insufficient to force the water all the way to the surface in one lift. Powerful electric pumps developed in the twentieth century have made it possible to cope with almost any amount of water from the pit bottom pump house.

As we have indicated, drainage has always been one of the most difficult and expensive of the coalowner's problems. Unfortunately, water has no respect for the boundaries which separate one pit's coal from the next. There have been countless disputes between colliery owners over drainage difficulties like those mentioned earlier in this chapter. Quite apart from water actually running from one mine to another, a sough or a powerful pump built to drain one mine will also help to drain neighbouring ground.

Sometimes colliery owners cooperated in erecting pumps or driving soughs, sharing the expenses in agreed proportions. One pump which was so built in the 1830s was the one at the old Calcutta Colliery near Thringstone, Leicestershire, whose massive engine house still stands. George Stephenson, who sank Snibston Colliery in 1831–2, initiated this joint drainage scheme. Calcutta was, by coincidence, the site of some of the earliest Newcomen engines in the East Midlands. These had been erected as long before as 1720. Whilst the Calcutta engine house still stands, the engine itself was dismantled after nationalization and replaced by an electric pump.

At another old mine in the same locality, California (Swannington) an engine was similarly erected to drain the Snibston and Whitwick Colliery reserves. In Shropshire, a Cornish engine at Wombridge and the Lloyds engine, Madely Wood, similarly drained several neighbouring mines. In some districts, the colliery owners established drainage commissions to run pumping stations, maintain soughs, and so on. This was so with the California pump. The two companies who had it erected formed a joint subsidiary, the Swannington Pumping Company, to run it.

Another drainage scheme, of much larger proportions, was undertaken to drain the Fitzwilliam Barnsley Bed workings in South Yorkshire. This scheme is still in operation as the South Yorkshire Mines Drainage Unit with headquarters at Rawmarsh. The Elsecar Pump installed in 1795 was part of the Fitzwilliam drainage scheme. The Unit maintains fourteen pumping stations, and many of the water levels can still be travelled.

Pipes have sometimes been used for draining coal works in much the same way as they are used for draining fields. A wooden pipe, about twelve inches square, was revealed at the Stead Mill 'B' opencast site to the south of Alfreton in the early months of 1969. Unfortunately, heavy rainfall caused the high wall to collapse before this pipe could be properly examined. Similar wooden pipes are sometimes used today mounted vertically in shafts.

In the small mines of Shropshire, clay pipes were often used for underground drainage. Examples have been found on several opencast sites (e.g. Princes End, Lawley) and some are now stored at Blists Hill. These pipes vary between about twenty and twenty-two inches in length and they taper from about 6¼ inches diameter at one end to 3¾ inches at the other; the clay is about half an inch thick. Obviously, they were slotted together. It is not known where exactly these pipes were made, but they are undoubtedly of local manufacture. From the workings in which they were found, it seems likely that they date from the late eighteenth century, but similar pipes have been used much more recently.[16] Similarly, in stall and pillar workings in the Top Hard seam at Coppice Colliery, Derbyshire, tiles were laid in the floor clay for drainage. Two of these, probably early nineteenth century, are now at Lound Hall.

Another device which has been used for drainage is the syphon. Indeed, a small drift mine at Alston, Cumberland, uses a plastic pipe syphoning system even today and there is also a syphon at Usworth Colliery, Durham.

CHAPTER SEVEN

Surface Arrangements

In the early eighteenth century, the surface arrangements at collieries were very simple. The author of the *Compleat Collier* tells us that the loaded corf was hooked on the rope in the pit bottom and drawn to the surface, 'where the Banck's-Man, or he that guides the Sledge-Horse, has an empty Sledge to set the Loaden Corfe on, as he takes it out of the Hook on the Pit-Rope, and then immediately hooking on an empty Corfe, he leads his Stead-Horse away with the Loaden Corfe, to what Place of the Coal heap he pleases'. One of the two banksmen, for a small addition to the normal wage, was charged with keeping check on the output of the various pitmen which he did with the aid of tally sticks.[1]

An early print of a coalmine at Chester-le-Street which appeared in the *London Magazine* in 1765 and is reproduced in Atkinson's book, shows a whim gin, a shaft which is fenced round, a banksman taking a corfe off the rope with the help of a long hook, a horse drawing a sledge and a heap of coal with four men sorting it. There are also two braziers whose purpose was to supply light because the colliery would be working during several hours of darkness in Winter. No doubt cannel coal would be used if it was available. It was unusual for a pit top to be securely fenced at this period. The sub-commissioner who visited Nottinghamshire for the Children's Employment Commission as late as 1841 saw only one pit in the district whose shaft was properly protected, and it was not unknown for a pit top worker to stumble or slip on ice and fall down the shaft.

A picture of a 'Coal Mine in the Midlands *circa* 1790' in the Walker Art Gallery, Liverpool, shows an atmospheric engine, two shafts with pulleys, a boiler, a chimney, a gantry and very little else. Similarly, the painting of Middleton Colliery, Leeds in 1814, to which we have referred earlier, shows a fairly modern looking headgear and close to it a small hut and three men one of whom is holding a horse; a steam winding engine with rope drum outside the engine house, a chimney, and a small heap of coal. There is nothing else on the heapstead (pit bank) but in the foreground is the colliery railway.

It would be interesting to know what happened to the bulk of the slack produced in northern coalmines in the eighteenth century. In the Midlands most of it was stowed in the mine, only the large coal and round coal normally being sent out. This is attested by the quantities of slack and small coal found in the wastes of old workings. On the Shirland site, the slack must have comprised something like a third of the total output. The workings there are most probably pillar and stall with total extraction because there are no signs of supports being set. Early longwall workings elsewhere in the Midlands show similar large quantities of slack in the gob. Lime burners were the only substantial users of slack from Midland coal mines until the late eighteenth century. Some mine owners in districts where there was limestone built their own kilns to absorb some of the make of slack. This was so at Skegby, Nottinghamshire, for example. Much of the lime used in the construction of the Elizabethan Hardwick Hall came from Skegby. Traces of the old lime quarries and works abound in the district.

In the late nineteenth century, and much of the early twentieth, colliers in the Midlands were required to fill coal with forks or screens instead of shovels. At an earlier period riddles or rakes were used to separate the saleable coal from the slack. Since the screening was done underground there were no great quantities of slack to be tipped on the surface. Where there was a market for slack, limited quantities were sent to bank separately from the lump coal and were paid for at a special price, usually about half of the piecework rate for coal.

Again, little dirt was made. Before about 1850, roadways were driven almost entirely in coal except that at some pits the main road was driven partly in stone to make extra height for pony haulage. Even after 1850 the quantities of dirt tipped on the surface were quite small. Colliers at many places in the Midlands were expected to 'gob' the dirt they made; and it was regarded as an offence for a stall man to send any out of the pit without permission. The result is that the dirt tips of collieries abandoned in the nineteenth century are small; and the tips of collieries abandoned earlier than the nineteenth century are almost non-existent. The size of the dirt tip is a good indication of the age of an abandoned mine.

To turn now to Northumberland and Durham, coal mining was, as we know, undertaken on a much larger scale than in the Midlands in the eighteenth century. True, powder was not used for coal getting anywhere in the north until about 1813 and was not widely used until the 1820s so that much less slack would be made than in districts where explosives were used. On the other hand the pressure of competition from new collieries forced producers of inferior coal to screen their product from about 1760. Dunn credited William Brown of Willington Colliery with introducing the first screening plant in 1740, but, as Galloway points out, this cannot be correct because this colliery was not sunk until 1775. The earliest screens were made of wood, but cast iron bars soon replaced the wooden ones.[2]

Again roadways were better in the north than in the Midlands to allow of pony haulage and to help with ventilation so that there was much more stonework. There was therefore more ripping dirt to dispose of; but whether this was stowed underground or sent out of the pit is not known. Early dirt tips in the north are certainly more in evidence than in other parts of the country, but they are still quite small by comparison with present day collieries, and much of the material in them could be accounted for by dirt wound out during sinking and heading out.

After the introduction of explosives for coal production in the north, small coal became a problem. It may be that up to that time the salt pans and the growing number of steam engines had absorbed the relatively modest tonnages of slack sent to bank. But this could hardly be true of the much larger amounts produced in the 1830s and '40s.

For example, at Pemberton Main Colliery, Monkwearmouth, sunk between 1826 and 1834 and costing between £80,000 and £100,000 to open out, the coal was discharged over three-quarter-inch screens through which about a quarter of it passed. Similarly at South Hetton, another large colliery, the coal passed over half-inch screens which separated out between a third and a quarter of the total product. In 1837 this one colliery sent over 204,000 tons of coal to London, so the quantities of slack being produced in the coalfield as a whole must have been enormous. It is most unlikely that a market was found for all of it. Many old dirt tips in the north can, therefore, be expected to have had a high proportion of combustible matter.

The process of banking and screening at northern collieries still using corves was described in Hair's *Sketches* of 1839 as follows:

> When the corves containing the coals are brought to the mouth of the shaft, they are dextorously unhooked from the rope by the 'banksman' whose business it is to place them on small hurdles, in which they are drawn to a range of 'screens' consisting of cast iron gratings, about half an inch asunder. Over these the coals are poured, for the purpose of separating the large from the small. The different kinds are then placed in 'waggons' each containing 53 cwt., shaped like an inverted prismoid, and provided with a brake, for retarding its speed when necessary. They are emptied by means of a trap-door at the bottom.[3]

However, while this description may have fitted the general practice, some of the actual sketches of surface arrangements are at variance with it. The sketch of the 'C' Pit, Hebburn Colliery, for example, shows three corves on the winding rope which have been lifted high in the headstocks, well above the normal decking level, to facilitate tipping directly over the screens. This practice, which eliminated the need to push the corves for some distance from the shaft side, became widespread in Northumberland and Durham.

By 1839 some northern collieries had, in any case, changed over to winding coal in wheeled trams lifted in cages. The first completely successful application of the new system, perfected by T. Y. Hall, was at Glebe Colliery, Woodside near Ryton in 1835. Twice as much coal could now be wound up in the same length of time and for this reason many other local colliery owners quickly replaced corves with trams and cages. True, Shilbottle Colliery near Alnwick still used corves in 1864 and William Colliery, Whitehaven used them until 1875.[4]

But most northern collieries adopted tubs and cages in the 1840s and even in the Midlands all the larger collieries had gone over to the new system by about 1860.

As we have seen, the screens in use in the northern coalfield in the first half of the nineteenth century were cast iron bars arranged like gratings. In the 1820s it was usual to have a primary screen to screen out coal over half or three-quarters of an inch; a secondary screen to screen out the 'peas and beans' of three-eighths to half an inch (or three-quarter-inch as the case may be) and a third screen to separate the fine coal below three-eighths of an inch, which was suitable for steam raising, from the dust which was unsaleable. The different grades, as they passed the screen, were delivered into carts or wagons down wooden chutes.

In the middle of the century, mechanical tipplers were used to tip the tubs over at the top of the screens. There are tipplers on a model of a pit top at the South Kensington Science Museum which represents a northern colliery in 1858.[5] Here there are double-decked cages with catches to hold the tubs firm whilst traversing the shaft. The cage is held steady at the pit top by manually operated wrought-iron 'keps'. The tubs, each holding about eight cwt of coal, are discharged over screens close to the shaft by tipplers mounted on trunnions. As a full tub is run on to the tippler, the latter overbalances, so tipping out the coal, and is then 'righted' manually. A tippler which was counterweighted so as to return tubs to a horizontal position automatically after unloading, was invented by James Rigg in 1870.

In the 1880s, jigging screens and picking belts came into general use in the north and at large collieries elsewhere. A jigging screen is a plate with holes of an appropriate size which is 'jigged' or shaken mechanically rather like a 'cake-walk'. A series of such screen plates separates the various sizes of coals.

The earliest picking belts were made from flat ropes stitched together, but these were soon superseded by iron plate belts. More recently, rubber belts have come into general use. As the mineral moves slowly along the belt, dirt is removed by hand. Until recently it was also the practice to hand select particular qualities of large coal. In some cases, these were 'dressed', that is, any bands of inferior coal or slatey matter were chipped off. Hand selected and dressed qualities were, of course, much more expensive than mechanically screened coals.

Perhaps it would be appropriate at this point to emphasize that early attention to coal screening was mainly confined to Northumberland and Durham where there was much fiercer competition between producers than elsewhere. In the first half of the nineteenth century little preparation of coal was done in the Midlands except by the collier at the coal face with his fork or riddle. However, in the second half of the century, coal preparation plants were introduced in all coalfields and the preparation processes became more involved.[6]

So far we have discussed only the dry processes. From mid-century coal washing plants came into use at many collieries. Coal washers sort out coal from dirt, and they can also separate different sizes of coal. They depend on the fact that heavier particles sink more readily in water than lighter ones. The dirt which is mixed with coal is, of course, very much heavier than coal. Various kinds of wash-boxes were invented in the nineteenth century. In all of them, movement was communicated to the water by manual or mechanical agitation or by compressed air. Such simple wash-boxes operate in much the same way as the gold prospector panning his dirt.

A more recent process for treating small coal, usually known as froth flotation, works on a different principle. Here an oily reagent is added to the water in small quantities and bubbles form when the mixture is agitated. These rise to the surface to form a fine froth. Small particles of coal are picked up by the bubbles and brought to the surface; whilst particles of dirt sink to the bottom.

Coal washeries use a great deal of water. However, they economize their use of water by having a clarification process. In the present century thickeners have come into use for this purpose. These are circular settling tanks with revolving arms or scrapers at the base. Polluted water from the wash-boxes enters the thickener and the solids settle at the bottom to be drawn off by the scrapers while fairly clear water runs off from the top of the tank.[7]

Coal preparation is a complicated subject and in a work of this kind it would be inappropriate to go into technical detail. The surprising thing which the layman will note, however, is that the principles underlying present day coal preparation practice have been known and used for so long. There have been many improvements in the equipment used but little fundamental change in techniques. Even the circular revolving picking table which has recently been adopted at some collieries in place of the picking belt is an old idea. Picking tables were used in French coal mines in the 1850s and were used in metal mines earlier still. Similarly, coal washing employs methods used for metals at an early period.

Screening plants are usually built on stilts, so that railway trucks can pass beneath them. Typically, wagons are hauled by locomotive from the main line up a slight incline to a point above the screens. There will be several roads passing beneath the screens depending on the number of qualities produced. The wagons are usually lowered through the screens by gravity, and the coal is fed into them down chutes. These are often adjustable booms which can be inclined into the wagons to minimize breakage.

The introduction of machines which both get coal and load it on to conveyor systems has altered radically the composition of the product as it comes up the pit (usually called 'the run of mine'). There is a much greater proportion of slack in the run of mine than was the case twenty years ago. Indeed, few collieries nowadays produce more than twenty per cent of large coal, and this is in comparatively small lumps. Again, power loader machines send a fair proportion of dirt out in the run of mine and this needs to be removed.

Consequently there is little really large coal to be hand selected, let alone dressed. Except at the dwindling minority of collieries which still fill a proportion of their coal by hand, there are few qualities over six inches in size. The best way to clean small coal is by washing, so that a high proportion of the product is washed nowadays. Washery plants are therefore much bigger than they used to be, while picking belts employ comparatively few people and they are concerned with removing large pieces of dirt rather than with selecting particular grades of coal.

Another allied feature of modern mining is that a higher proportion of the run of mine—sometimes as much as a half—finds its way on to the dirt tip. Disposing of slurry from the washery is a particularly difficult, and growing, problem.

The industrial archaeologist will find few traces of disused screening plants or washeries. Engines and engine houses, fans and headstocks sometimes survive, but screening plants do not. They are large utilitarian structures which have no alternative uses; and there is little of the romantic about them. The only old screening plants which survive are those still in use at working collieries, and even these are fast disappearing. Hand screens (i.e. forks) used formerly in the Midlands are, however, occasionally found in old workings; and they are still used in many coal depots where coal is bagged by hand.

On the other hand, dirt tips do survive. Early ones are, by now, pleasant grassy mounds frequently covered by trees and shrubs. Later ones need no description. However, the National Coal Board in conjunction with local authorities, is improving the appearance of many modern dirt tips. The contours are rounded and, in appropriate cases, grass is sown by a new and rather expensive process. Youth clubs have cooperated by planting trees.

All colliery dirt tips contain some coal and because of this they are liable to spontaneous combustion. The resultant red shale is useful for road building and similar purposes. Consequently, some tips have provided aggregate for motorways; others have been, and are being, quarried by building contractors and others.

In the mining valleys of South Wales, dirt disposal has always been much more of a problem than elsewhere, because there is so little flat land available. A nineteenth-century colliery might find room for its sinking dirt in the pit yard, but soon it would have to start tipping on the hillside. Particular attention is nowadays given to the stability of such tips.

So far, we have described various surface features—headstocks, fans, engine houses, screens, washeries, railway lines—but have said nothing about workshops and stores.

All collieries use large quantities of materials of various kinds: roof supports, explosives, rails, belting, conveyor structure, to name but a few. Occasionally an explosives store, a strongly constructed building, will be left standing when a colliery closes, even where the other buildings are completely obliterated. The reason for this is that the main explosives store is always built at a distance from the pit head. Bulky materials, such as pit props, are kept out in the open and leave no trace once the colliery has closed.

The workshops are important buildings, housing mechanics, electricians, blacksmiths, joiners, painters, and others. The earliest craftsmen employed regularly about the works were probably the corvers. As J.C. explains in the *Compleat Collier* of 1708:

> And whereas I speak of Corves or Baskets to put the Coals in, we must have a Man (which is called the Corver) to make them.
>
> He must have a good Quantity of young Hazle Rods, provided for that Purpose, with young Plants, or Sippleings, as we here call them, of Oak, Ash or Aller, of about three Inches thick, or better, for the Corf-Bow; we buy the Rods by Bunch each Bunch, containing about a Hundred Rods, at about Six-pence *per* Bunch, and the Bows being better than two Yards long, for half a Crown or three Shillings *per* dozen or thereabouts.
>
> Your Corver ought to be just to you, in keeping up your Corfe,

> for with Working the Coals, being drawn pretty briskly up, the Corves are subject to Clash and beat against the Shaft sides, and so beats down your Corfe dayly, that if your Corves be not dayly beat up and mended; you may lose more than one Inch dayly, which would bring your Measure or Corfe, of fourteen or fifteen Pecks, down to nine or ten Pecks, and so you lose a third of your Measure.[8]

At Blundell's Collieries, Lancashire, the hazel rods from which the corves were made were made pliable by being put in a cylinder into which steam was blown, and no doubt this, or something like it, was done elsewhere. The corver was replaced in the second half of the nineteenth century by the tram (or tub) repairer, often known as the 'tub thumper'. Originally, he was a carpenter since tubs were made of wood; in the present century, steel trams gradually replaced wooden ones, calling for rather different skills.

Another early craftsman, the enginewright, was much more highly skilled than the corver. He looked after the pumps and other machinery; and with the introduction of steam the enginewright developed into a civil and mechanical engineer although often retaining the old title. Some colliery enginewrights in the nineteenth century were responsible for the introduction or development of new equipment and techniques. The most famous of these was, of course, George Stephenson, who was the chief enginewright for several collieries on Tyneside belonging to Lord Ravensworth and partners when he invented his first locomotive. His brother, Robert Stephenson senior was also a colliery enginewright before becoming a manager. Another whose achievements were more modest but still important was John King, an enginewright at Pinxton Collieries, Derbyshire. His most famous invention was a safety detaching hook which we have mentioned already, but he made several other contributions to the improvement of winding apparatus. And for every John King whose inventions became widely known, there were a hundred other enginewrights who succeeded in making important modifications at their own collieries which were not, perhaps, of general application.

The blacksmith in the nineteenth century was chiefly responsible for the shaft and its equipment besides the more usual blacksmithing work (e.g. shoeing horses). Pick sharpening was usually done not by the colliery blacksmith but by a specialist having his own separate forge. He was usually paid by the men since they were responsible for finding their own tools and keeping them in repair. Until recently most old collieries had a primitive pick-sharpener's shop with hand-bellows but most of these have now gone. One of the exhibits at the Lound Hall Mining Training Centre is a leather bellows from the blacksmiths' shop at Brinsley Colliery.

One of the carpenter's most important jobs was to make, and keep in repair, ventilation doors. Some of the bricklayer's work was also to do with ventilation; making airtight stoppings at vulnerable points.

Among other specialists were boiler smiths and locomotive fitters. There are comparatively few steam locomotives used at collieries nowadays but many of the old loco sheds survive at mines which are still open.

Certain new surface features appeared at many collieries towards the end of the nineteenth century or in the early years of the twentieth, for example, air compressors and power houses. From about 1925 pithead baths and canteens began to make their appearance. The disposal of large amounts of dirt has necessitated aerial ropeways and conveyors at many places, although in recent years dumper trucks and bulldozers have partly replaced them. As we shall see when we come to discuss coal face mechanization, Alfreton Colliery, Derbyshire, had three air compressors and receivers in 1898 and these provided power for the mine's three coal cutters. This colliery's surface plant may also be used to exemplify the changes in boiler house equipment. For this comparatively small mine there were eight Lancashire boilers; six were twenty-eight feet six inches by seven feet and the other two were thirty feet by seven feet. It is also interesting that five of Alfreton's boilers were fired by gas produced at the Colliery by a Wilson's patent producer size eight feet by ten feet. The other three boilers were fired by hand, and the steam pressure was eighty pounds per square inch.

As we have seen, coal preparation plants also underwent changes in response to changes in the marketing situation and the quality of the product.

Basically, however, the typical colliery surface of 1947 appeared much as it had in 1870. Steel headstocks may or may not have replaced wooden ones, but this made little difference to the general appearance. The boiler-house with its range of hand fired boilers; the winding engines; the heapstead; the poky stores with its banks of wooden bins holding nuts and bolts and suchlike, drums of oil, bags of cotton waste; the blacksmiths' shop sometimes still having hand bellows; the fitters' and carpenters' shops where hand tools were still used to make parts for machinery and equipment of various kinds; the stables for the horses usually with the provender store above, these and many more features showed little change. There are few of these old style collieries left now. Is it too much to hope that one of the few remaining ones will be converted into a living museum of the mining industry when its time for closure comes?

A detailed treatment of associated activities like agriculture, brick-making and the production of coke and bye-products, would be out of place in a volume on coalmining. Briefly, coalmining as it was carried on by landed proprietors in the nineteenth century and earlier was regarded as part of the general estate economy; and the connection between farming and mining has since been cemented by the need for horse-feed and for land for tipping. Brickmaking provided a use for small coal and for the clay and shale often associated with coal seams. Coke manufacture was a natural development of colliery enterprise. The early methods of coking, in the open air and then in bee-hive ovens, wasted the valuable bye-products which are so important a feature of modern coke ovens.

CHAPTER EIGHT

Coal Face Mechanization

Before the Industrial Revolution the scale of operations at a colliery was limited by shaft capacity, and by drainage and ventilation problems. With the wide adoption of the Newcomen engine for pumping, about the middle of the eighteenth century, drainage was no longer a limiting factor. It was now possible to lay dry all the seams which a colliery proprietor might wish to exploit. Similarly, furnace ventilation, dating from the second half of the eighteenth century, was adequate (by the standards of the time) for mines of the then optimum size. Steam winding, which was adopted widely from about 1800, increased shaft capacity although this was still a limiting factor until the introduction of the cage held steady in the shaft by guide rods in the 1840s.

Underground haulage was then the limiting factor at many collieries. Boys on haulage commonly worked longer hours than men at the coal face. This was necessary because it took longer to transport the coal to the pit bottom than to win it at the face; and as faces moved further out from the shaft the problem worsened. Double shift coal getting would have imposed an intolerable strain on the haulage system in most districts; though it was common in the Northern coalfield where hours of work at the coal face were particularly short. Here, the youths and boys on haulage worked considerably longer hours than face men. The introduction first of steam, and later of compressed air and electricity, to drive haulage engines underground, largely corrected this situation.

By the end of the nineteenth century, the main limiting factor was clearly to be found at the coal face. Coal face work had altered very little in two hundred years. The only substantial change in technique was the introduction of explosives in the early nineteenth century. The substitution of steel hand tools for wood; and the more systematic setting of timber to support the roof must also have had a marginal effect on productivity; but on the other hand, the safety lamp tended to reduce productivity as compared with the candle because it gave such a poor light.

Various attempts had been made to lighten the work load of the collier by designing mechanical coal-cutters to replace hand-holing. Holing (i.e. undercutting the seam) was undoubtedly the most laborious job at the coal face. The holer had to chip away at the solid coal with a heavy pick, lying for much of the time on his side and then on his back to get under the cut. Incidentally, many holers suffered a deformation of the hands, the middle finger becoming permanently bent inwards towards the palm. Cutting mechanically would not only reduce the labour cost of holing but also free a good pick man for work as a coal getter.

The early coal-cutting machines imitated the process of hand-holing. Instead of one man wielding the pick directly, the pick was mounted in a frame and operated through gearing by one or two men rotating a crank handle. William Brown, a well-known Northumbrian coal viewer, designed such a machine in 1768. This was known as 'Willie Brown's Iron Man'. Many similar machines, operated by hand, were tried out over the succeeding seventy or eighty years. A horse-powered 'iron man' was also tried at a Yorkshire colliery in the early nineteenth century.

However, such machines were unsuccessful, being less efficient than the hand holer with his pick. An effective coal-cutting machine needed power to drive it. Some attempts were made to apply steam power to coal cutting equipment through chain or rope drives, but the equipment was cumbersome and dangerous. The earliest satisfactory power source for face machinery was compressed air, first used underground at a colliery near Glasgow in 1853.

Inventors sought to apply this new supply of power to versions of the 'iron man'. One of these inventors, Thomas Harrison, patented such a power-driven swinging arm machine in 1863, but followed it a few weeks later with a greatly improved design embodying a horizontal disc which undercut the coal in the manner of a circular saw. This was not a new idea, but it was the first successful application of the circular saw principle to coal-cutting.

Several other people designed disc machines in the 1860s and 1870s. Perhaps the most widely used of these early coal-cutters was the Winstanley. One of these was put to work in the McNaught

20 (*above*) Rope wheel at Clifton. 21 (*left*) Use of the hand-drill at Clifton Colliery

22 Early A.B. disc coal-cutter at Anderson-Boyes' works

23 A.B. Meco-Moore cutter loader

Seam at Barleith Colliery, Ayrshire in 1874. The owners of this and neighbouring collieries were John Galloway and Co., who were pioneers of coal-cutting machinery. William Galloway, the colliery manager, subsequently became a Professor of Mining at Cardiff. Another coal-cutting machine, installed by the same firm in 1873, worked on a different principle. This was the Gartsherrie machine, built by William Baird and Co. of Glasgow. The Gartsherrie employed a jib (a projecting horizontal arm) which carried around its edge an endless chain in which cutter picks were fixed at intervals.[1]

The Gartsherrie coal cutter was the first practical chain machine. One was exhibited in 1876 at Philadelphia, U.S.A., and American manufacturers subsequently produced greatly improved machines employing the same principles. In Britain, meantime, disc coal-cutters were found to be more robust than the early chain machines. One disc machine thought to be particularly robust was Clarke and Stevenson's. Three of these machines were at work at Alfreton Colliery, Derbyshire, in 1898 when a party from the Midlands Branch of the National Association of Colliery Managers visited the works. These machines were at work in the Deep Hard Seam. They cut in the hard clunch at the base of the seam taking out a kerf four and a half to five inches in thickness to a depth of four feet three inches. The discs were five feet in diameter and they were made in two pieces so as to be easier to transport in confined spaces below ground. Each machine had two cylinders of nine inch diameter and a stroke of nine inches. The gearing was nineteen to one and the revolutions per minute between 200 and 250.

These machines were driven by compressed air. There were three compressors at the surface. Two were Ingersoll-Sergeant Straight Line Compressors, Class 'A', but were not of equal size. One had an air cylinder eighteen and a quarter inches in diameter, a steam cylinder of sixteen inch diameter and a one foot six inch stroke; the other had an air cylinder twenty-four inches in diameter, a steam cylinder twenty-two inches in diameter and a two foot stroke. The third compressor was a different make (Sturgeons) having an air cylinder twenty inches in diameter, a steam cylinder eighteen and a half inches in diameter and a stroke of two feet six inches.

These three compressors delivered air at forty-five to fifty pounds per square inch into three receivers which were near the engine house. The air was conveyed down the shaft in ten inch cast iron pipes which fed it into a receiver near the upcast pit bottom. From this receiver the air travelled up the main roads in cast iron pipes of six inch and four inch diameter and then along the gates to the coal faces in three-inch pipes. Finally, it was carried along the coal faces to the machines in two-inch diameter hose pipes, which were about thirty yards long.

Clarke and Stevenson also made an electrically driven disc machine from about 1893. Like the compressed air cutters used at Alfreton, these machines were strongly constructed and could be used to cut into a hard floor.

However, disc machines had some disadvantages. For example, the large disc could not follow the contours of an undulating seam. In such a seam it was necessary, therefore, either to cut in the floor (which might be impossible if the floor was hard) or cut well up into the coal. Since cutting machines in these early days were used chiefly in thin seams where labour costs were high, cutting above the floor was something to avoid. Another drawback of this type of coal-cutter was that it could easily become fast in the cut. The reason for this is that there was such a large expanse of disc under the seam all the time that, if the coal was fragile, it would drop on to the disc and trap it.

A third type of coal-cutter, the bar machine, facilitated the insertion of chocks of wood in the cut immediately behind the cutting member, thus reducing the risk of jamming. The cutting arm on this type of machine was a tapered round steel bar, armed with cutter picks along its length. The picks were arranged in the form of a spiral which helped to expel slack coal from the cut as the bar rotated. One such machine, dating from the late 1870s, also had a spiral groove to bring out the slack. This innovation of Warsop and Hill has had a wide application in recent years.

Another early bar machine patented by Bower and Blackburn in 1887, was the first coal-cutter to be driven by an electric motor. This was of ten horse-power. Electric motors were subsequently used for many disc and chain machines too.

The bar type of coal cutter was not nearly so robust as the disc machine and breakdowns were consequently frequent. Similarly, when American machinery manufacturers reintroduced machines of the chain type in the early years of the twentieth century, they were found to be much less reliable than disc machines. These chain machines worked reasonably well in America because the coals are generally softer there than in Britain. Quite a few such coal-cutters were sold to British colliery owners because they were offered on very attractive terms, but they proved a disappointment. It was not until after 1910 that improved chain machines were able to challenge effectively the more traditional disc machines and not until the 1930s did they establish their superiority. In 1911 some 471 British mines (of a total of about 3,100) used coal cutters. The total number of cutting machines in use was 2,146 of which 1,020 were disc machines, 390 were bar machines and 148 were chain machines. The remaining 588 were mainly of the percussive type. Some 1,148 of these machines used compressed air and the remaining 998 were driven by electricity.

The introduction of electricity underground was obviously hazardous and regulations having statutory force were imposed in 1905. Subsequently, safety requirements were made more demanding. In particular, all electrical equipment for use underground had to be made flameproof so as to minimize the risk of explosions.

The percussive machines referred to were for development and short wall work. The most popular percussive machine was the Siskol, known in some places as the 'Pom-pom' because of the sound it made in use. The Siskol Company always specialized in drills, and their cutter had much in common with a drill. It could be adjusted either to undercut or overcut the coal, or to cut it vertically. In undercutting, the drill struck the coal face repeated blows in quick succession whilst being moved sideways in an arc. It could then be used to shear the coal vertically in circumstances where the use of explosives to bring down the coal was inadvisable. A similar electric heading machine employing a three h.p. or five h.p. motor was also made by Siskol. In this case, the cutting bit operated by rotary motion and not by percussion; and the machine was mounted on wheels, whereas the compressed air machine was not. One Siskol cutter, dating from about 1913, was recovered a few years ago from the Rock Mine, Ketley, by the Shropshire Mining Club. It is now exhibited at the Shrewsbury Borough Library and Museum. This machine is 'complete with drills, angling gear, starter, cable, glands, service dan, servicing book and chart'.[2] Similar machines were widely used as late as the 1940s.

Coal-cutting machines were adopted much more readily in

Scotland than in most English counties.[3] Even as late as 1938 some eighty per cent of Scottish coal was cut by machine as against fifty-six per cent for England and Wales. The probable reason for this is that wages costs at the coal face were comparatively high in Scotland and so there was a greater inducement to spend capital on labour-saving equipment. An additional factor is that hand-holing was more difficult and proportionately more wasteful of coal in the thin seams of Scotland than in the comparatively thick seams of some English counties. It should be understood that the kerf on machine-cut faces was much thinner than on hand-cut faces, and the amount of slack produced by the cutting process was, therefore, lower. This might not be too important in a six-foot seam; but it certainly was so in a seam only two or three feet thick.

As we have seen, the disc machine was found more suitable for British conditions than the chain machines introduced from the U.S.A. from about 1902. However, the American firms offered their machines on such favourable terms that British manufacturers were concerned. Consequently, attention was given to the development of improved chain coal-cutters to compete with the American products. One of the earliest of these improved British machines was a chain cutter first marketed by Anderson Boyes of Motherwell in 1906, although it did not sell very well and was soon withdrawn. This machine was twenty and a half inches high, thirty-five inches wide and eight feet nine inches long, and was electrically driven. It was simple in design and much more robust than contemporary American machines. It was mounted on skids and the height of the cut could be varied.

Some colliery companies were more persistent in the early use of coal-cutters than others. There was always a temptation to go back to hand-holing where cutting machines had proved to be inadequate; or where (as often happened) industrial relations difficulties followed the introduction of machines. One firm which persevered with coal-cutters was Blundells of Lancashire.[4] This firm installed 'Diamond' disc coal-cutters in 1902. The Diamond, invented by W. E. Garforth, was a popular machine in the inland coalfields. One at Blundell's collieries, compressed-air driven, cut 3,014 yards in a month which was a remarkable performance in the early years of the century. In 1907 the firm bought some bar machines made by Mavor and Coulson of Glasgow. Five of these machines were used at Blundell's

Pemberton Colliery on a semicircular coal face a mile in length. The same colliery owners also purchased percussive machines made by Siskol and two other firms, Hardy and Patterson. In 1920 they tried a Jeffrey chain machine but were apparently not impressed with it. In 1929 they had nine bar machines, thirty-four percussive machines and forty-four compressed air drills, but no chain machines.

During the 1920s and 30s the chain coal-cutter gradually gained ground as designs improved, and by the beginning of the Second World War there were comparatively few other types in use. In the case of Blundell's collieries, they had eight chain machines in 1937 against only one bar machine; although there were still fifteen percussive machines in use.

Pneumatic drills were never as popular in Britain as in some continental countries, but they were adopted in some coalfields. They were especially useful where the presence of gas made shot-firing inadvisable, or where faulting or an uneven floor precluded the use of conventional coal-cutting equipment. In 1938 they were used at 485 collieries out of a total 2,125 collieries in Great Britain. Some 1,685 pneumatic drills were used with cutting machines, and another 7,250 were used independently for coal getting. At the same date, British mines used 140 disc machines, 186 bar machines, 6,005 chain machines and 1,398 percussive machines, a total of 7,729 coal-cutters.

It is regrettable that very few early coal-cutting machines have survived. Because they are convenient for breaking down, they are almost invariably sent for scrap when no longer required. A few have found their way into museums, including the Siskol machine already mentioned; and one or two are at Coal Board workshops. For example, the N.C.B. Northumberland Area have a Black Star coal-cutter dating from about 1920 which was unearthed in old workings a short time ago.

The second step in coal face mechanization was the face conveyor. Before the introduction of face conveyors, longwall faces had cross gates at fairly short distances apart to facilitate the transport of coal to the main gate. These cross gates absorbed much labour not only in ripping but also in packing dirt to support the roof. An alternative method was to cast all the coal by hand along the face, or to draw it along in small tubs (where there was sufficient height) or on flat, wheel-less boxes called dannies or jotties. The face conveyor reduced the number of roads to be driven to a minimum; and similarly eliminated the necessity of casting coal down the face. But it created some roof control problems which were not insuperable but which nevertheless discouraged many owners from trying or persevering with the new system.

One of the first face conveyors, the Blackett, was invented in 1902. It consisted of a trough running along the length of the face through which ran an endless scraper chain. The fillers cast the coal into the trough, and the scraper chain drew it to the main gate to be loaded into trams. A second type of conveyor was invented by Thomson, a Scottish colliery manager, and first manufactured by Anderson Boyes in 1908. It 'consisted of a train of trays jointed together and extending a little less than half the length of the face; it could be worked from either end. When loaded, the train was drawn by a haulage gear across the centre gate, where a plough scraped the load from the trays into the tubs, the operation being repeated each time the trays were filled.'[5]

Another type, originating on the Continent, was the shaker conveyor, used mainly where the face dipped towards the loader gate. This consisted of a series of steel pans (or troughs) to which a jigging motion was conveyed by a compressed air or electric motor, the coal being shaken from one pan to the next. In the 1920s the men often called this type of conveyor a 'jazz' because of its motion.

In the 1930s, another type of conveyor became pre-eminent. This was the band (or belt) conveyor; an endless conveyor belt running on rollers and driven by pulleys from the loader gate end. It was pioneered by Richard Sutcliffe and Company.

The face conveyor carried the coal along the face, but it still had to be loaded into trams at the gate end. One way of doing this was to have a sunken roadway allowing the tram to run under the end of the conveyor. The other, and more usual, method was to have a mechanical gate-end loader.

By 1938 there were some 7,826 conveyors (mainly face conveyors) and 766 gate-end loaders at work in British mines against 3,218 and 355 respectively in 1929. The quantity of coal mechanically conveyed had risen from 37,150,000 in 1929 to 122,915,000 in 1938.

The rubber conveyor belt was, by 1947, regarded as part of the conventional longwall system of mining. Since 1951, other materials like PVC have been used instead of rubber so as to lower the fire risk.

At this point, it might be useful to summarize the cycle of operations on a typical longwall coal face of the period. On one shift the coal was undercut by machine (usually, by this time, electrically driven). Wooden chock nogs were inserted in the cut to give temporary support to the coal. On the next shift, holes were bored along the face at intervals preparatory to shotfiring; the shots were fired thus loosening the coal, and the fillers (otherwise known as 'colliers' or 'coal getters') got to work. They levered the coal down, broke it into manageable pieces and shovelled it onto the conveyor belt. Having cleared their fifteen or twenty tons, they made the face safe for the third shift. On this third shift the flitters (or 'erectors') moved the conveyor belt forward to the new face line whilst the packers built packs in strips along the line of the gob from which the coal had just been removed.

This conventional longwall system was, however, under challenge from a new system where the coal was both cut and loaded by machine. One of the first power loaders to be used on a coal face was put to work at Easthouses Colliery, Scotland, in 1924. The machine employed was an American one with a ranging drum 'which brought the coal down into a gathering conveyor as the machine advanced'. Another early attempt at power loading was made in the 1920s in Cannock Chase with a 'Jeffrey flight conveyor which was jacked into cut-and-prepared coal'.[6]

But the first effective longwall power loading machine was the Meco-Moore. The early Meco-Moore cutter loaders, manufactured in the 1930s by the Mining Engineering Company of Worcester, combined what were in effect two separate machines, a cutter and a loading unit. First the coal was cut as the machine moved in one direction (the 'cutting run') and then the coal was blasted down. Next, the machine returned along the face and its loading unit loaded the prepared coal on to the face conveyor. During the war years, this machine was vastly improved as the result of sensible cooperation between its originators, the Anderson Boyes Company, and the Bolsover Colliery Company. Now, machines were designed which cut and loaded simultaneously instead of having two separate operations. The early experiments were carried out at the Bolsover Company's Rufford Colliery, then managed by the present Production Member of the National Coal Board, W. V. Sheppard. Continuous improvements were made and by the end of the war, the A.B.

Meco-Moore Cutter-Loader was a fully proven machine. Its action has been described as follows:

> The A.B. Meco-Moore Cutter-Loader does three jobs in one. As it travels along the face it cuts the coal, collects it and moves it to a conveyor running parallel to the direction of the traverse. Its action resembles that of a harvesting machine in so far as the coal is first cut and then delivered to a conveyor which forms part of the machine. The cutting portion consists of an A.B. Fifteen Special Longwall Coal-cutter with two horizontal jibs, one cutting at floor level, the other at a point near the middle of the seam. In addition, there is the shearing jib which cuts from roof to floor at the back of the web. Pressure of the roof breaks down the coal which either falls into the trough-like loader attached to the back of the machine, or is delivered on the endless slatted loader belt by a 'riffle' bar fitted with a number of picks which break up the bottom coal if it has not already collapsed. The loader belt delivers the coal sideways to the face-conveyor which is laid hard against the roof props. The loader apparatus is reversible so that the machine can travel back along the face for the next cut. Fitted to the side of the delivery end of the loading jib is a device to sweep up the gummings—coal dust—which are produced by the cutting jibs and accumulated during operations.[7]

This machine required a team of six or seven men. In addition, it was necessary to have a 'stable hole' got out by hand at each end of the face to provide the machine with room to turn round, and these absorbed a similar number of men. However, these thirteen or fifteen men replaced the thirty or so coal getters required on a hand-filling face. By the end of 1948, some seventy of these machines had been produced and about forty or fifty were in use.

The Meco-Moore was a cyclic machine. It replaced the coal getters, but people were still needed on another shift to move the conveyor over and to build packs along the face.

A second type of power loading machine was the plough which was developed in Germany during the war and it is still extensively used there. The plough is essentially a plane which is hauled up and down the coal face like an endless chain ploughing the coal off in both directions. Unlike the Meco-Moore it is a non-cyclic machine, but operates successfully only where there is a strong roof and floor in combination with soft coal.

After the war, several makes of plough were introduced into British mines with varying success. Similar in many ways were the Huwood Slicer-loader and Mavor and Coulson's Samson Stripper. Like all power-loading machines, these work well in some conditions but not in others.

During the later stages of the war and in the early postwar years, many British mining engineers went over to America to examine modern coal getting equipment of various kinds. With more enthusiasm than judgment many of them subsequently introduced American equipment to British mines. The results were often lamentable because the equipment was not suited to the conditions. The American power-loading machines were designed for stall-and-pillar work (there was, at that time, hardly any longwall work in the United States) and consequently there was a temporary rash of stall-and-pillar in districts which had for generations worked nothing but longwall. However, this phase soon passed.

The American experience was by no means completely wasted and some of their equipment, like the scraper-chain conveyor, came into regular use here. However, substantial progress in mechanization could only come through improved longwall power-loading machinery designed for British conditions. In fact, since 1950 there has been a whole new generation of power loading machinery and its associated equipment. Almost all the new power loaders are either shearers or trepanners of one kind or another.

The Anderton Shearer was designed to run mounted on an Armoured Flexible Conveyor (A.F.C.), a device pioneered in Germany. The A.F.C., as its name implies, is a robust steel structure.

A shearer is similar in principle to a longwall coal cutter. However, where the cutter merely undercuts the seam, the shearer has a horizontally pivotted drum on which several rows of picks are mounted. The diameter of the drum is roughly the same as the height of the seam so that, as the machine goes forward, a complete section of coal is sheared off and gathered up by a plough attached to the machine which forces the coal into the A.F.C. (Fig. 15).

Unlike the shearer, the A.B. Trepanner is a floor-mounted machine which runs alongside the A.F.C. The trepanner head is

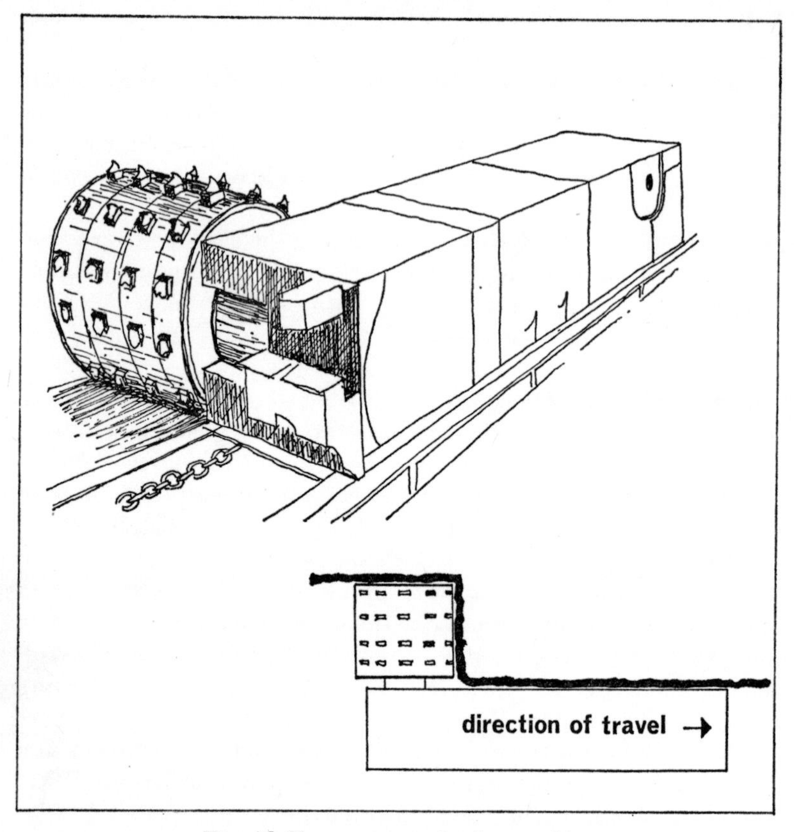

Fig. 15 Shearer power-loader machine

like a giant auger with cutter picks fixed round its periphery. As the trepanner moves forward, the trepanner head bites into the coal, again taking off a complete section (Fig. 16).

There have been many variants of both the shearer and the trepanner. For example, there is a bi-directional shearer which (as its name implies) cuts in both directions. An important recent innovation is the double-ended conveyor mounted trepanner. Yet another power loader, the trepan-shearer, combines features of both types of machine. All these machines are non-cyclic. No preparatory work is necessary as was the case with conventional mining and even the Meco-Moore. Coal production can go on through all three shifts.

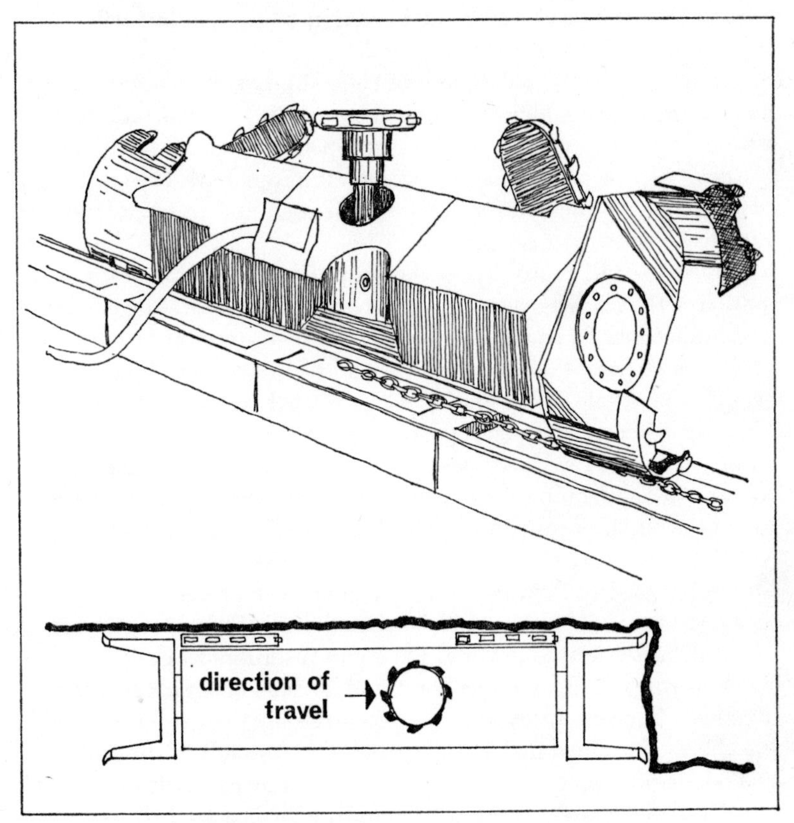

Fig. 16 Trepanner power-loader machine

Improved systems of support have been a great help. With conventional mining, the roof was supported by rigid props and bars (either wood or steel) at the working face. After the coal had been won, packs were built in strips along the face into the waste (gob). This was a necessary part of the process of roof control.

The first postwar improvement was the hydraulic prop which operates in much the same way as the hydraulic jacks used in garages. The miner sets the prop vertically, and then by the operation of a key 'pumps' the crown of the prop up to the roof where it exerts an immediate pressure of five tons. As the roof settles down, the prop will resist pressures of up to 20 tons without yielding. Over twenty tons the prop will yield a fraction of an inch but will maintain its upward thrust undiminished.

Whereas with the old system of rigid supports it was necessary to have a line of props along the face side, cantilever bars could be used with hydraulic props to provide a prop free front to facilitate the running of the power-loader. Hydraulic props have the additional advantage that they can quickly be released and re-set as the face advances.

There are still many hydraulic props in use, but for mechanized coal faces they have been largely superseded by hydraulic chocks. A hydraulic chock is, in effect, a structure which combines from two to six hydraulic props (called 'legs') with one or two cantilever bars or beams, the whole being mounted on a steel platform to form one unit. Attached to the platform is a hydraulic ram. These hydraulic chocks are set a few feet apart along the length of the face. As the power loader machine moves along the face, cutting and loading a strip of coal, the rams on the chocks are activated first to push the A.F.C. forward to the new face line and then to pull the chocks themselves bodily forward. This pushing over process goes on continuously throughout the shift so that there is little roof exposed at the coal face at any one time. On some installations each group of chocks is controlled from one point and a whole batch moves forward together. There is an even later system, so far little used, where all the chocks are controlled from a panel (or 'console') in the gate.

The degree of roof control made possible by hydraulic chock installations obviates the need for building face packs. Instead, as the chocks move forward, the 'gob' is allowed to cave in completely. Thus, there is very little hard physical labour along the length of the face; but the picture at the face ends is rather different. First, the machine needs room to turn round at each end of the face, and so it is necessary to hand fill sufficient coal to provide a 'stable hole'. True, stable hole machines are used in some places and various ways of eliminating the tail gate stable hole have been proved. For the time being, however, this is still primarily a manual job in most districts.

The gates are still, for the most part, ripped by hand. The dirt is blown down by explosives, then packed at the gate side (or loaded out). Then steel arches, lagged with wood or corrugated iron sheets, are set as permanent roadway supports. Even here, effective machines are being developed for ripping in tail gates and some are in use. The newest ripping machine, the 'boom' ripper pioneered in the North Nottinghamshire Area is without doubt much the best and its use may well become universal within the next two or three years. However, there is still only one way to build a really solid gate side pack; and that is by hand.

Nowadays, power loading and ripping machines are fitted with water sprays to keep down the dust. This helps to minimize the incidence of chest complaints, but it brings its own problems.

For the industrial archaeologist, the speed of technological change today poses a problem. Whilst relics of early mining practice survive, relatively modern machines are scrapped. For example, a pick found in the Hutton Seam at Burnhope Colliery, in workings earlier than 1742, and now at Graham House, Newcastle-upon-Tyne is in excellent condition. On the other hand, it seems that not a single example of the Meco-Moore power loading machine has survived in any condition, despite the fact that quite a number were still in use five or six years ago.

CHAPTER NINE

The Miner and the Mining Community

Coal mining in the Middle Ages was carried out mainly by peasants. It was, indeed, regarded as a branch of agriculture. Even in the first half of the nineteenth century the collieries of the landed aristocracy and gentry, like the Middletons of Wollaton, were under the superintendence of the bailiff along with the farms and mills and cottages. Further, with the exception of the North-East coalfield, coal mining was no more than a part-time occupation before the nineteenth century. There was, therefore, nothing very special about the style of dress of the early coal miner. So far as we know he dressed for the pit much as he dressed for the farm.

By the eighteenth century, coal mining had become, for many, a full-time occupation in Northumberland and Durham. There may, therefore, have been in these counties something special about the style of dress as compared with farm labourers. Leifchild, writing in 1856, gives this description:

> Their pit dress is made entirely of coarse flannel and consists of a long jacket with large side-pockets, a waistcoat, a flannel shirt, a pair of short drawers, and a pair of stout trousers worn over them. Add to these a pair of 'hoggers' or footless worsted stockings, a tight-fitting round leather cap, and you have the hewer ready for the pit. A pit suit costs about one pound, if purchased at the slop-shop; though some wives and daughters can make them.[1]

Flannel suits are nowadays worn only by sinkers and shaftsmen, and occasionally by others working in particularly cold and damp conditions. The reference to the leather cap is puzzling. Leather caps were worn by colliery officials (overmen, master shifters, deputies, etc.) and also by sinkers. They were worn, as an illustration in Atkinson's book shows, with the peak at the back. But they were not normally worn by ordinary pitmen. A few of these leather caps have survived and there is one at the Northern Regional Open Air Museum.[2]

The northern colliery official of the nineteenth century was distinguished not only by his leather cap but also by his blue flannel suit, his yard-stick (for measuring work) and his pocket watch.[3] Boys wore much the same sorts of clothes as the men. In 1837 when George Parkinson started work down the pit at the age of nine years, he 'donned the pit-boy's flannels, and with the "bait poke" over my shoulder and the candle-box in my pocket, I looked down on the poor boys who had to continue at school'.

George Walker painted a picture of a Yorkshire miner, at Middleton Colliery, Leeds, in 1814. He has a muffler round his neck, an old black hat with brim turned down all round, and cloth leggings. The editor of the book in which Walker's painting was reproduced said that Yorkshire miners wore white cloth suits bound with red,[4] but it is difficult to believe that they went to work in them even if they were frequently washed. The suit worn by the man in the picture looks far more like a walking out suit than a working suit, and this is indicated by the walking stick he carries in one hand and the basket over his free arm.

According to a Children's Employment Sub-Commissioner,[5] at many pits in the southern part of the West Riding, men in 1841 worked 'in a state of perfect nakedness' whilst the girls of six to twenty-one years of age who assisted them were 'quite naked down to the waist'. The same Sub-Commissioner reminds us of another item of equipment worn by haulage hands in low seams:

> One of the most disgusting sights I have ever seen was that of young females dressed like boys in trousers crawling on all fours with belts round their waists and chains passing between their legs, at day pits [i.e. drifts] at Hunshelf Bank, and in many small pits near Holmfrith and New Mills; it exists also in several other places.

In Somerset the place of the leather dog-belt was taken by the guss, a harness made of hempen rope, still in use in the early years of the twentieth century. One such guss, worn about fifty years ago, is now preserved at the Lound Hall Mining Training Centre.

The walking out dress of the Durham miner in mid-century was solemn black—on Sundays at least—and this was usually attributed to the Methodist revival. Previously pitmen are said to have worn loud and colourful clothing in their spare time. They were certainly not short of clothes. When George Parkinson went up to the chapel on Saturday evenings to prepare for his Sunday school class, he wore his 'second best' suit, so he must have had at least three.

In the later nineteenth century, colliers in general wore old suits at work just as they do today, with thick woollen stockings, heavy boots, and cloth caps. Many Durham miners had the trouser legs of their pit suits cut short at about knee height, in order to be able to pull the trousers on and off with their boots on, according to Atkinson. Officials still wore their blue flannel suits.

Pit helmets were not introduced until the early 1930s, but they are now universally worn. They carry a bracket on which the cap lamp fits. Knee pads, again, have only become widely used in the last thirty or forty years. Previously, colliers in low seams often knelt on boards. Leather safety gloves have also become fairly popular in recent years, but many men refuse to wear them because they are thought to be cumbersome.

To the extent that miners were drawn from the general agricultural population, their houses, like their clothes, were in no way special. Miners in Kingswood Chase were apparently expected to build their own hovels as late as 1675.[6] We may well imagine that they were crude wooden structures, possibly worse than those of the average farm labourer, rather than better. The cottages of the rural poor were still, in the main, built of timber in the seventeenth century, or half timber, although by 1675 bricks were cheap enough at any rate to facilitate the erection of chimneys.

The eighteenth century saw the building of mining villages in many coalfields. In the Midlands the majority of miners might still live cheek-by-jowl with their neighbours in other occupations, but in Northumberland, Durham and Scotland most of them lived in isolated pit villages.

The houses in these villages were built in rows. They were small insanitary structures. In Scotland many of them were single-roomed hovels, with roughly built walls about five feet high, bare earth floors and no foundations. Some of these were still occupied in the 1860s and probably later.[7]

In Durham and Northumberland standards were a little higher. There were some one-roomed cottages there, too, but most had

24 (*above*) Hopkinson 'Black Star' coal-cutter at Ashington Colliery.
25 (*left*) Koepe winding gear at Ashington Colliery

26 Teversal Colliery pit top *c.* 1915

27 Teversal Colliery Plowright tipplers (removed *c.* 1930)

either one or two ground floor rooms and a loft. A sketch by S. H. Grimm (1778) shows a two-storied row of pitmen's dwellings with outside staircases. It is clear that the rooms in the upper storey are separate from those in the lower, and it may be that each family occupied only one room.[8]

In most cases, the rows were built in pairs, with their fronts facing each other. The narrow space between the fronts was generally clean. Where there were several pairs of rows, their back doors faced each other across a wider space. As Leifchild says:

> The space between each two rows of back doors, presents along the centre one long ash heap and dung-hill—generally the playground of the children in summer, with a coal heap and often a pigsty at the side of each door. Each row generally has a large oven, common to all its occupants; there are no conveniences.[9]

Incredible as it may seem, many of the eighteenth- and early nineteenth-century houses commented on so unfavourably by Leifchild and others over a hundred years ago were still occupied in the 1930s, indeed, quite a lot survived into the 1950s. A study of miners' housing was undertaken in the 1920s by R. A. Scott-James in connection with a Liberal Party enquiry under the chairmanship of David Lloyd George. The report of this enquiry contains some astonishing photographs taken at the time—which could just as well have been taken in the mid-nineteenth century.

Indeed, the Report shows that in many cases the original one-roomed stone hovels of County Durham had been enlarged to accommodate a growth in the mining population by the addition of a second storey. This may account for the two-storied dwellings, as sketched by Grimm in 1778.

In the meantime running water had at least been made available and this represented an improvement over wells, but only a very slight one because, in many cases, the water had to be drawn from stand-pipes, of which there were one or two for each row of houses. Slops were still carried away down an open drain at the backs of the houses. Earth closets were now provided in wooden or brick structures in the yards of the houses, but these were shared and many people preferred still to use them as little as possible.

Scott-James entered a house at Consett with one room downstairs and another, entered by a ladder, in the rafters, in which a man, wife and six children (of whom three were over eighteen) lived.

There were plenty of similar examples in Scotland, too. As Scott-James pointed out, in Lanarkshire, there were 61,202 houses with single rooms and 155,285 with two rooms of the county's total of 321,471 houses. Many had no back doors or windows, some indeed were built back-to-back; in some, coal had to be kept under the bed; in many, water had to be brought in from stand-pipes outside.

Bad housing conditions were found in South Wales and Derbyshire, too, but nothing like so bad as the worst examples from Scotland and the north-east.[10] The oldest slums in Scotland had been built whilst Scottish miners and their families were still serfs tied to their employer's coalworks as surely as the medieval villain was tied to his lord's manor. Serfdom was not abolished in Scotland until 1799. In Northumberland and Durham the men were engaged under yearly bonds until the 1840s and this tied them also to their employers although not nearly so strongly as serfdom. Their houses were provided either rent free or at a nominal charge. It seems that the servile status of the Scottish collier families was reflected in the atrocious hovels built for them in the eighteenth century. This cannot be said quite so strongly of Northumberland and Durham.

Even in the three counties referred to, however, miners' housing was not universally bad. The houses at New Lambton, a Durham mining village dating from the late eighteenth century, were regarded with nostalgia by George Parkinson who was born in one in 1828. Like many others, the house in which he was born had only two rooms, a general room downstairs and a loft entered by a ladder, but there was a pantry also. The house place had a brick floor. His grandmother's house in the same village was similar, but with two rooms downstairs. The picture Parkinson draws is almost idyllic: 'One long row of low-roofed brick cottages, with a few other rows standing apart, formed the street, which faced a meadow through which ran a clear burn or stream.' He then speaks of an old water mill, delightful woods, and primroses growing in spring, 'so that even in the life of a pit village, beauty and variety of interest were not lacking'. As to the interior of the houses, he tells us about one occasion when:

> The little table was set in front of the window, covered with a 'harn' tablecloth. Everything, though rough and coarse, was made

> spotlessly clean. The fireplace, bright with polished fire-irons and a glowing blaze, shone welcome. Behind the door a ladder led to the upper room or loft close to the tiles, which were not hidden by any plaster or wooden ceiling. The flooring boards of the loft were laid loose upon the joists.

The impression one gets here is of houses small but solidly built, and adequate by the standards of the time. At any rate, mother and father could sleep in a separate room.

Many old miners' rows in the north-east, improved at various times, still stand. One improvement was the provision of a cold water tap either in the back kitchen or the pantry. Hot water was provided by a boiler at the side of the fire. Black-leaded kitchen ranges, with a boiler at one side of the fire and oven at the other, became standard in miners' houses in the second half of the nineteenth century.

Toilet facilities have also been improved. Many houses originally built without lavatories, had privies added later. At first these were earth closets. Often, as at Chopwell, the ashes from the kitchen were tipped in the midden, so that when the 'midden man' came to clear out his task was rather less unpleasant than it would otherwise have been. These earth closets have in their turn been replaced by water-closets within the last twenty or thirty years.

There are still some one-bedroomed cottages left similar to Parkinson's old home except that they have stairs in place of the ladders common in his day. Downstairs, they still have the one room (called the kitchen by older people, but the living room by others) plus the pantry with its tap. Two- and three-bedroomed terraced houses for the larger families have a back kitchen as well as a pantry.

Later miners' rows were much the same in all coalfields. One especially well-built row in Nottinghamshire which survived until recently was Portland Row, Selston, erected in 1823. Here, there were forty-seven brick-built houses in a continuous row with no alleyway between any of the houses. The first house in the row was bigger than the rest, being built as a 'tommy shop' and ale house. (Miners at this date were paid partly in kind—the so-called Truck System and the 'tommy shop' was run by the colliery owner or someone with whom he had an arrangement.) The other houses were of a type which was to become familiar in all colliery districts 'two up and two down'.

In a typical dwelling of this type, there was a back kitchen (or scullery) with a cold water tap over a flat sink and a coal-fired copper for boiling clothes; a kitchen (or living room) with a black-leaded kitchen range having a boiler on one side of the fire-grate and an oven on the other. In some parts of Durham and Northumberland, it was traditional to have a stew-pot, replenished daily, always on the hob. The only other downstairs room was the pantry, usually built under the stairs. The two bedrooms would measure somewhere between ten feet and twelve feet square as a rule and at least one of them would have a fireplace. Outside there would be a coal-house and closet. In many rural districts these were either earth or tub closets until after the Second World War. In a minority of cases instead of a copper in the kitchen there was a separate wash-house with a copper at the back of the house.

These houses were usually built in rows; but sometimes they were built in squares instead. Two good examples were Napoleon Square and Holden Square at Cinderhill, Nottinghamshire, built in the early 1840s but these have also been demolished recently.

Of course, there were some larger houses of the same general type. These 'double' houses usually had a small scullery, a kitchen and a parlour downstairs; two bedrooms on the first floor; and an attic built in the slope of the roof on the second floor.

An essential piece of equipment for all these houses was the zinc bath which was usually kept hanging on a nail outside the back of the house. On bath night the bath was put on the hearth rug and filled with hot water from the boiler. One lot of water, topped up as it grew cool, did for the whole family. Some miners took a hot bath in this way every day, but many contented themselves at most, with a strip-wash at the kitchen sink or in a large bowl. The difficulty of ensuring privacy for a nightly bath must be appreciated, especially where there were several young children.

There are plenty of houses of this type still occupied. Later miners' houses, built in the 1920s and after, resemble Council houses of the same period. The Butterley Colliery Company was altogether in advance of its time, however, in building miners' houses with bathrooms at Kirkby-in-Ashfield, Nottinghamshire, as far back as the late 1880s.

The public house has always been of central importance in the mining village. Coal mining is thirsty work. Besides, in the early nineteenth century there was no other secular building in which communal activities could be carried on.

We have noted that No. 1 Portland Row was larger than the other houses in the Row because it was built as a beer house and tommy shop. Judging by an out-building with square chimney, the beer was home-brewed. Further along the Row, two houses (Nos 25 and 26) were, at the time of the 1851 Census, occupied as one tenancy described also as an ale-house. As we shall see, this tavern had another, seemingly incompatible, function at the same time. Here, then, are examples of public houses built as part of a miners' row which survived until recently. Similar examples no doubt still survive in various coalfields. Such places, while differing little from the houses of the miners, offered a bit more room, a chance to smoke a pipe, to talk, to drink, to listen to the newspaper being read, to play dominoes or some other game, to gamble and to get away from the family for an hour or two.

Some other public houses used by the miners were more substantial. Plenty of eighteenth- and nineteenth-century examples survive although most have been tarted up in recent years. There can be few colliers' pubs with whitewashed walls and sawdust on the floor nowadays.

One good local example of an old colliers' public house recently modernized is the Broad Oak at Strelley, a village where coal was mined from the later Middle Ages until early this century. Externally this fairly small brick-built pub has retained much of its original appearance except that its outdoor games facilities have been sacrificed for a car park. But indoors the place has been 'improved' out of all recognition. A Leicestershire example which has not been altered to anything like the same extent is the George Inn, Peggs Green. There has, again, been coal mined in the vicinity for some hundreds of years.

Of course, these Midlands public houses were used by agricultural labourers and framework knitters as well as by colliers. The public houses built along with the pitmen's rows in Northumberland and Durham are more truly colliers' pubs. Parkinson had this to say about the one at Lambton in the early nineteenth century:

The only place for social gatherings or recreation was a public-house, formed by uniting two cottages, which with a fenced cockpit and a quoit ground at the front, and a quiet place for pitch-and-toss just round the corner, provided opportunities for votaries of these sports, which, with the tap-room as their centre, were often accompanied by drunken brawls and fighting, with all the demoralizing influences arising therefrom.

An example of a pitman's public house which is part of a terrace is at Washington old town. A good Midlands example is the 'All Nations' in Coalport Road, Madeley, Shropshire, where the ale is still home-brewed. Several Shropshire public houses have coal seams outcropping in their cellars, the best example being the 'Pheasant' at Broseley. Similarly at the 'Peacock Inn', at Oakerthorpe, Derbyshire, coal has actually been worked by a drift leading from the cellar. One public house in Durham, the 'Happy Wanderer', Framwellgate, has a small museum of mining artifacts. There are many 'Miners' Arms' in all coalfields. One example, near Castleford, Yorkshire, has a weather vane showing a miner wielding a pick-axe instead of the more usual weather-cock.

The tavern at 25 and 26 Portland Row, Selston, in the mid-nineteenth century was interesting because upstairs it had a room which was used as a chapel by the Original Methodists. It seems that an extension was added at the rear of these two cottages shortly after the Row was erected in 1823, the ground floor being used as an ale house and tommy shop; and the first floor as a chapel seating seventy-five people in discomfort. Shortly before the Row was demolished, Donald Grundy took photographs some of which appear in the *Proceedings of the Wesley Historical Society* for June 1968. Part of the original structure of the chapel had, however, disappeared some years earlier.

Similarly, the Methodist Chapel at Lambton which was built in the 1830s, was a converted miners' cottage, this time at the end of a row. Parkinson tells us that: 'the colliery workmen were sent, and part of the needful timber was provided. The roof was raised several feet, partitions removed, and a gallery at each side and at the back was put up; two large windows put in, and the doorway protected by a small porch.' It appears that: 'this structure, being somewhat higher than any other building in the place, had a prominence which naturally attracted attention where there was little else to notice'.

Many early nineteenth-century chapels were subsequently converted to other uses as their congregations grew. One at Huthwaite, near the Nottinghamshire–Derbyshire border, was sold in the early years of this century and was then converted into a grocery shop with an off-licence. A clue to the original use of the building is provided by an upstairs window of ecclesiastical shape. This chapel was replaced by a larger and more pretentious structure which is still in use. In more recent years many chapels have disappeared or been converted to other uses not because their congregations were too large; but rather that they were far too small.

Many of the early miners' welfare institutes and miners' union offices were built with a distinctly chapel-like appearance. The Leicestershire Miners' Offices at Coalville may be taken as a surviving example. Many leaders of the miners in the nineteenth century and early twentieth were Methodists and this no doubt accounts for the similarity.

In recent years many interesting old miners' houses and other buildings have been demolished and others, now empty, await a similar fate. It is important that surviving examples should be adequately recorded. At the time of writing, a party of students from the Welsh School of Architecture, under the supervision of Jeremy Lowe, are about to make measured drawings of unoccupied N.C.B. houses at Engine Row, Mechanics Row and Stack Square, Blaenavon. Similarly, E. A. Dixon, the N.C.B.'s Area Estates Manager in the South Midlands Area has had sketch plans made of The Cob, a small terrace of stone-built back-to-back houses at Moira, Leicestershire, built about 1811, which are about to be demolished; and three houses at Church Street, Church Gresley dating from 1711, which will also be demolished when they are empty. Plans showing old houses in the Barnsley Area (Top Row, Woolley Colliery and Long Row, Carlton Terrace) have been furnished by the Estates Manager for that Area, J. M. Bellis. The houses at Portland Row, Selston, mentioned earlier, were sketched by scholars from a local grammar school under the supervision of their then history master, Mr Stevenson. It is to be hoped that more of these properties can be similarly recorded before it is too late.

Mention might also be made of the Wesley Historical Society's plan to record historically interesting Methodist Chapels, classified according to whether they are still used as chapels, or have been converted to other uses, or have been demolished.

Graveyards are a useful, if melancholy, source of industrial archaeological evidence. For example, at Madeley Church, Shropshire is a gravestone erected to the memory of the victims of a shaft accident. Similarly, Annesley cemetery, Nottinghamshire, has a gravestone giving the names of men dying in a minor explosion, whilst the 204 victims of the Hartley disaster of 1862 are commemorated on a memorial in Earsdon churchyard.

CHAPTER TEN

Conclusion

The geographical distribution of Britain's coal industry changed to a remarkably small extent until recently. Kent is our only new coalfield; production there did not commence until 1907. The other main coalfields have been exploited since the late Middle Ages. They are: Scotland; Northumberland and Durham; Yorkshire, Derbyshire and Nottinghamshire; Leicestershire and South Derbyshire; Staffordshire, Salop, Worcestershire and Warwickshire; South Wales and Monmouth. In addition, we may mention four small colliery districts which are historically important: Cumberland; Westmorland; Gloucester (the Forest of Dean) and Somerset (around Radstock and Bristol). In addition, small quantities of coal have been mined in Dorset and Devon.

There was a gradual shift in location as coal on the outcrop became worked out, but the closure of whole districts is a mid-twentieth-century phenomenon. There are now no deep mines in Worcestershire, South Staffordshire or the Forest of Dean, and only one in Shropshire. In 1947 these were all thriving colliery districts. Similarly, the older parts of Scotland, Northumberland and Durham now have few mines; and soon even the Rhondda valleys will cease to be a coal producing area.

This rapidity of change presents local industrial archaeology groups with a challenge. Detailed surveys of collieries and their associated railways and secondary industries (e.g. the manufacture of coke and bricks) need to be undertaken before it is too late.

Magnificent work is being done by the North East Industrial Archaeology Society on the initiative of the Durham I.A. Group whose secretary, Don Wilcock, has been in communication with me. Surveys are being made of collieries (both working and closed), mining villages, and tram roads. A copy of their recording sheet for closed collieries is given as Appendix A.

The Manchester Region Industrial Archaeology Society carried out a coal mines survey in August 1968. The results were summarized in the Group's *Newsletter* No. 1 of Spring 1969, a copy of which was supplied to me by their secretary, A. D. George. Various members of this group are extending the survey. For example, Reg Schofield and John Smethurst are surveying the tramways of the Irwell Valley before they are obliterated by improvements.

Brief reports of work done by other groups appear from time to time in the David and Charles periodical *Industrial Archaeology*. These groups concern themselves with surface features. But the importance of a coalmine stems from what lies underground. Internally, the National Coal Board holds an enormous quantity of historically interesting material. There is a recognized procedure for dealing with books of account and similar documentary material, much of which is deposited at County Record Offices. Plans of old workings are held in the Board's Regional and Area offices where they can be consulted by serious students.

N.C.B. Areas also hold photographs and drawings of plant and equipment, underground workings, and so on; but there is no uniform system for recording these and there is no central catalogue. It is feared that many drawings of historical interest but of no practical use have been, and continue to be, discarded. This is understandable: drawings accumulate; many are ephemeral; but among them will be a minority which ought to be permanently retained. But who is to distinguish between them when the time for a general 'sort out' comes? The criterion normally used for determining whether or not to keep a drawing is: has it any practical value today? A drawing of a fan at a closed mine may be of interest to an industrial archaeologist, but if the fan has been scrapped the chances are that the drawing will be discarded. This is the sort of problem to which the proposed Mining Historical Society will no doubt be able to draw attention.

While groups of industrial archaeologists will normally concern themselves with surface features, they should keep in mind the possibility of surface indications of shallow underground workings, particularly where old shafts are known to be. Occasionally, they may find it possible to enter old workings but they would be unwise to do so without skilled guidance. It is all very well for groups like the Shropshire Mining Club who know what they are about; although even for them some danger exists.

Old workings exposed on opencast sites involve less danger, although even here care needs to be exercised. An opencast contractor in the Midlands who entered an old roadway exposed on a site which he was working was overcome by blackdamp and lost his life. Falls of rock from the high wall also occur, so it is as well to wear a safety helmet. It should also be remembered that opencast sites and artifacts found on them are private property. Groups wishing to visit a site should therefore obtain proper authority in advance; and should not remove artifacts without the permission of a senior official.

Coal mining has experienced two periods of rapid technological change: 1840 to 1860 and 1947 to the present day. Before 1840 change was a gradual process. The greatest single technological innovation was the steam engine, adopted for pumping at deep mines in the eighteenth century, and for winding at most collieries between 1790 and 1830. These early winding engines were very small and crude, however, and the speed of winding was little greater than with the whim-gin.

Between 1840 and 1860 the cage held steady in the shaft by guide rails replaced the corve swinging on a loose rope. It was now possible to bring coal on rails in one container from the coal face to the screens. In the Midlands furnace ventilation was adopted widely for the first time, thus making possible a considerable extension in the scale of operations. Larger and more powerful winding engines and pumps were installed at many places; and mechanical fans were introduced by the more progressive companies. Larger collieries installed mechanical haulages underground; and everywhere roadways were made to higher standards.

The second great period of change has been even more dramatic. In 1947 most work at the coal face was done by hand. There might be a cutting machine to undercut the face; and a belt to convey coal to the gate end although some collieries lacked even these. But nowadays almost all coal is won by power loading machines and many pits have roadway ripping machines too. Hydraulic chocks provide a self-advancing canopy of steel in place of the indifferent support of props and bars and strip packs. The average coal face today costs about £200,000 to equip, rather more than the average colliery was worth in 1947. A modest output for such a coal face is 1,000 to 1,200 tons a day equivalent to the output of a fair sized colliery of 1947.

On the underground roadways, large mine cars (often hauled by locomotives) have replaced small trams. In other cases trunk conveyors bring the coal most of the way to the pit bottom. Many steam winding engines have been replaced by electric engines; and large skips have in many cases replaced cages in the shafts. Steam locomotives, universal in 1947, have been replaced by diesel locomotives in most parts of the country.

But the greatest change is the reduction in the number of collieries managed by the N.C.B. from 980 in 1947 to 300 today. When a colliery closes, it disappears all too soon. For example, Clifton Colliery, the subject of some photographs taken underground which appear in this volume, closed in 1969. Now, the site is completely bare. There is therefore some urgency in recording everything of interest at collieries whose life is uncertain. Because of the likelihood of changes in capital equipment, it is also important to record things like steam winding engines and wooden headstocks at long-life collieries.

For winding engines, the sort of information required is indicated by this example kindly supplied by my colleague W. Humphreys:

East Wales Area
Elliott Colliery; East pit
Winding Engine
Type: Horizontal, double compound
Drum: Bi-cylindro Conical
Date of installation: 1891
Maker: Thornewill and Wareham of Burton-on-Trent
Dia. of drum: 15 to 25 feet
Drum speed: 55 r.p.m.
Rope speed: 72 feet per second
Winding rope: $1\frac{1}{2}$ inch diameter, flattened strand
Steam pressure: 130 lb per square inch
Winding depth: 526 yards

Together with photographs, an engineering drawing (if available) and indicator diagram, this gives a fair record. It will be appreciated that things like winding engines are so large and cumbersome that few can be saved from the scrap merchant. This makes it all the more important that the essential data should be recorded.

Coalmining was the foundation on which Britain's industrial

greatness was built. In the age of steam, coal was the only substantial primary fuel. Now it has competitors. But coal will continue to supply a high proportion of Britain's energy needs for many years to come. It follows, therefore, that even the most sophisticated equipment in use today will be superseded by even better equipment. Before long today's latest power loader machine will be a fit study for the industrial archaeologist. Unfortunately, such is the pace of change, machinery is often discarded and scrapped before its future value as material for the industrial archaeologist is recognized. Many recent machines and items of equipment have thus disappeared. If an adequate record of mining's technological development is to be maintained, those industrial archaeologists with an interest in this field need to exercise constant vigilance. It is all too easy to recognize the importance of a particular piece of equipment when every example has been scrapped.

APPENDIX A *Recording Sheet used by Durham I.A. Group*

NAME		LOCATION		O.S.	CARD NO.
NICKNAMES		1ST SUNK		RECORDING GROUP	C.R. CARDS
ORIGINAL OWNERS		LEASED FROM		FINAL CLOSURE	
SUBSEQUENT OWNERS				REASON	
Shaft names	Details	Seams worked	Methods	Intermediate closures	Reopenings
Ancillaries		Special features			
Transportation		Major reference sources			
Sketch on reverse		Original deposited			

Old Winding Engine—Glyn Colliery

This engine is reported to have been made by the Neath Abbey Works. There is no date on the engine, but as the pumping engine house is dated 1845, it is reasonable to conclude that the winder was installed earlier.

It is a single cylinder, double acting, vertical steam winder, with thirty inch bore and five foot stroke. The cylinder is below floor level and operated vertically upwards, driving two reel drums. Each reel is fifteen feet diameter, and carried $4\frac{1}{2}$ by $\frac{7}{8}$ inch flat winding ropes.

The left hand reel is of cast iron spider construction and the right hand wheel is made of substantial elm wood construction and bolted to a cast iron brake wheel.

A wood lined strap brake was in use, but dead weighting and a donkey brake engine were added later.

The valve motion is for hand operation only, and according to hearsay the winding enginemen had very good control of the engine by this method. It has even been said that the control was so sensitive that they could crack a nut between the cage and pit bottom when landing.

The engine was in regular use up to 1932.

Winding conditions

A very unusual feature of the arrangement, was that winding was carried out in two shafts simultaneously, through the downcast shaft twelve feet diameter, and the upcast shaft nine feet diameter. Single cages were used in each shaft and each carried one tram. The winding depth was 180 yards approximately.

The original headframe has been removed, but from foundation etc., on site, the construction appears to have been of the type as shown on the sketch.

Winding engine house

The house is made of dressed block stone and is a fine example of the good class workmanship of that period.

The roof is of the hip type with slates laid on substantial wood roof members.

N.B. See sketch (Fig. 17).

(Contributed by W. Humphreys, Area Chief Engineer, National Coal Board, East Wales Area.)

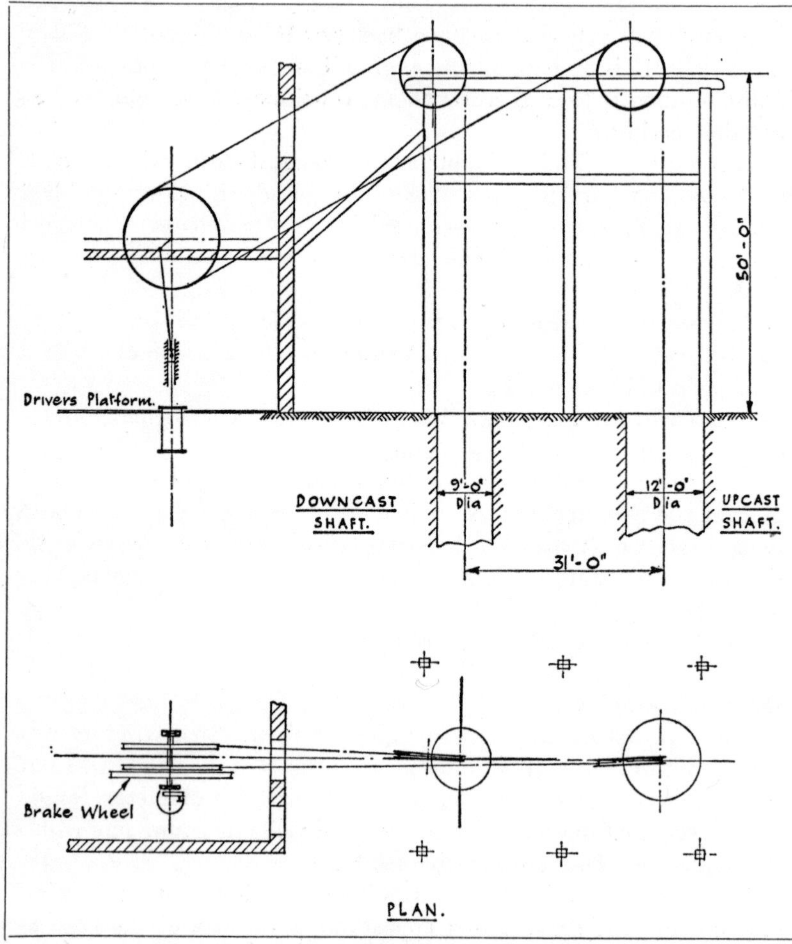

Fig. 17 Winding arrangement at Glyn Colliery

28 Little Eaton tram road

29 Single-storey houses, Stone Row, Leasingthorne

30 Fire bucket and hand winch, Rock Mine, Ketley, Shropshire, 1965

31 Strikers working on the outcrop, Shropshire, 1912. Note the bicycle wheel used for winding

Old Pumping Arrangements—Glyn Colliery

The beam engine and pumps were installed by the Neath Abbey Engineering Works, in 1845.

It is a vertical, double acting, single cylinder engine, with twenty-four inch bore and six foot stroke, and operated with steam at fifty pounds. The condensing pump is operated from the main beam. The piston rod is linked to an overhead oscillating beam which in turn is coupled to a crank shaft carrying a seventeen foot diameter flywheel. A suitable speed reduction is obtained through a pair of massive cast iron toothed gear wheels of four inch pitch.

The second motion shaft has a crank on one end coupled to a long horizontal wood connecting rod actuating a bell crank lever system. The extreme end of the lever carries the spear rod for operating the pumps in the shaft and to the opposing end of the lever is suspended a large mass of dead weights to counterbalance the weights of spear rods etc.

Pumping was carried out in two stages giving a delivery of nine to 12,000 gallons per hour with a total static head of 300 feet. The shaft depth is 186 yards, the pumps delivering into a water course some eighty-five yards from the surface and free flow taking place through a heading which delivered the water to a lower surface level.

The first stage of pumping was obtained by coupling a spear rod to the main rod above the water course previously mentioned. To the end of the rod is attached a nine inch pitch pine ram working in a cast iron cylinder, the latter being coupled to a simple suction and delivery valve bar, and delivery from this stage taking place through a twelve inch rising main, and effective for a height of 100 feet.

At this point a bucket piston operated in the rising main and motion was given by an extension of the main spear rod.

On the downward stroke of the piston a valve or valves in the piston permitted the water in the column to flow past the piston and on the upward stroke the water was lifted and delivered into the water course via a launder.

(Contributed by W. Humphreys, Area Chief Engineer, National Coal Board, East Wales Area.)

Source References

CHAPTER ONE *Early Mining Methods*

1 R. G. Collingwood and J. N. L.Myres, *Roman Britain and the English Settlements*, (2nd edn Oxford University Press, 1937), reprn 1963, pp. 231–2.
J. U. Nef, *The Rise of the British Coal Industry*, London, 2 vols, 1932, i, pp. 1–2.
I. A. Richmond, *Roman Britain*, Penguin, 1954, pp. 159–60.
2 Nef (i, 20) suggests that output rose from about 210,000 tons in 1551–60 to near three million tons in 1681–90 and over ten million in 1781–90, but these can be no more than very rough approximations. Output in 1913 was over 287 million tons.
3 J. Butt, *The Industrial Archaeology of Scotland*, David and Charles, 1968.
4 F. A. Henson and R. S. Smith, 'Detecting early coal workings from the air', *Colliery Engineering*, June 1955, p. 256.
5 F. Atkinson, *The Great Northern Coalfield* 1700–1900, University Tutorial Press, 1968, p. 11.
6 *Ibid.*; also *Historical Review of Coal Mining*, London, 1925, p. 48 (subsequently cited as *Historical Review*).

CHAPTER TWO *Transport*

1 Nef, *The Rise of the British Coal Industry*, i, p. 84.
2 *Ibid*, i, pp. 79 and 95.
3 J. C., *The Compleat Collier* (London, 1708), Newcastle upon Tyne, Northern reprints no. 4, p. 47.
4 A Wollaton rook was probably equal to about twenty-five cwt.
5 R. S. Smith, 'Huntingdon Beaumont: adventurer in coalmines', *Renaissance and Modern Studies*, i, 1957, p. 121.
6 R. S. Smith, 'Britain's First Rails: a reconsideration', *Ibid*, iv,

1960, pp. 123–31, also I. J. Brown, *The Coalbrookdale Coalfield, Catalogue of Mines*, Shropshire County Library, 1968, p. 35.

7 Cited R. L. Galloway, *Annals of Coalmining*, London, 1898, p. 156.

8 *Ibid*, p. 249.

9 *Ibid*, p. 250.
This method was still used at a colliery near Stella until the late nineteenth century.

10 *Ibid*, pp. 224 and 203.

11 R. L. Galloway, *The Steam Engine*, London, 1881, pp. 213–16.

12 R. Abbott, 'The railways of the Leicester Navigation Company', *Trans. Leics. Arch. Soc.*, 1955, pp. 51–61.
C. P. Griffin, 'The Economic and Social Development of the Leicestershire and South Derbyshire Coalfield', unpublished Ph.D. thesis, Nottingham, 1969. I wish to acknowledge the extensive use I have made of this thesis.

13 F. Atkinson, *The Great Northern Coalfield*, p. 52.

14 Galloway, *Annals of Coalmining*, p. 278.

15 *Ibid*, p. 324.

16 *Ibid*, p. 325, also *Children's Employment Commission (Mines) 1842, First Report, App. pt. ii*, p. 288 (evidence of J. Knighton).

17 C. Hadfield, *The Canal Age*, David & Charles, 1969, pp. 28–30.

18 T. H. Hair, *Sketches of the Coal Mines in Northumberland and Durham*, London, 1839, p. 14.

19 Galloway, *The Steam Engine*, pp. 227–9.

CHAPTER THREE *Shafts and Winding*

1 W. Gray, *Chorographia: or a Survey of Newcastle-upon-Tyne* (1649), 1883 edn, p. 20.

2 J. C., *The Compleat Collier*, p. 12.

3 D. Anderson, 'Blundell's Collieries: technical developments 1776–1960', *Trans. Historical Soc. of Lancs. and Cheshire*, vol. 119, 1967.

4 Galloway, *Annals of Coalmining*, p. 320, also Anderson, *loc. cit.*

5 J. E. McCutcheon, *The Hartley Colliery Disaster, 1862*, Seaham, E. McCutcheon, 1963, p. 49.

6 Report by J. K. Breakwell, cited *ibid*, pp. 165–70.

7 T. J. Raybould, 'The development of organisation of Lord

Dudley's mineral estates', *Economic History Review*, Dec. 1968, p. 536.

8 Galloway, *Annals of Coalmining*, pp. 297–301.

9 A. R. Griffin, *Mining in the East Midlands, 1550–1946*, Frank Cass, 1971.

10 *Proceedings of the Institution of Mechanical Engineers*, Oct.–Dec. 1903.

In addition to the Watt and Newcomen engines mention should also be made of an improved atmospheric engine, the Heslop engine, which probably infringed Watt's patent. This had two cylinders, one at either end of the beam. One was called the hot cylinder and the other, which acted as a separate condenser, was called the cold cylinder. This engine was little used outside Lancashire and Heslop's native Shropshire. However, one such engine erected at Howgill Colliery, Cumberland, in 1793 was taken to the South Kensington Science Museum upon going out of use—see Galloway, *Annals of Coalmining*, p. 355. There were three Heslop engines in use at Madeley Wood in 1789.

11 I am indebted to C. Buchan for this information.

12 G. Parkinson, *True Stories of Durham Pit Life*, London, 1911, pp. 17–18.

13 Galloway, *Annals of Coalmining*, pp. 481–4.

14 I am indebted to I. J. Brown for information on winding, and much else, in Shropshire.

15 *Industrial Archaeology*, Feb. 1969, pp. 98–9 and information supplied by the Brora Colliery management via R. Storer.

16 R. Wilson, 'The Koepe patent system of winding at the Bestwood Collieries, near Nottingham', *Trans. of the Chesterfield and Derbyshire Institute of Mining, Civil and Mechanical Engineers*, 1883.

CHAPTER FOUR *Winning and Working*

1 His expression is '3 yards or better, according to the Strength or Softness of the Coal'—J.C., *The Compleat Collier*, p. 43.

2 *Ibid*, p. 42.

3 *Historical Review*, pp. 82–8.

4 *Ibid*, pp. 48–54.

5 *Ibid*, p. 44.

6 J. C., *op. cit.*, p. 30.
7 D. Anderson, 'Blundell's Collieries: technical developments, 1776–1960', *Trans. Hist. Soc. of Lancs. and Cheshire*, vol. 119, 1967, pp. 123–4.
8 George Owen, cited in Galloway, *Annals of Coalmining*, p. 120.
9 See e.g. *Historical Review*, p. 95.
Atkinson's date (1841) is clearly inconsistent with the facts. Mechanical haulages had been used on inclines some twenty years earlier.
10 *Children's Employment Commission (Mines) 1842, First Report, App. Pt. ii*, p. 253 (Report of J. M. Fellows).
11 Anderson, *loc. cit.*, p. 131.
12 Galloway, *Annals of Coalmining*, p. 185.
13 *Historical Review*, pp. 95–6.
14 Anderson, *loc. cit.*, p. 133.
15 *Historical Review*, p. 199.

CHAPTER FIVE *Ventilation*

1 *Historical Review*, p. 127; Galloway, *Annals of Coalmining*, p. 254.
2 *Historical Review*, p. 113; Galloway, *Annals of Coalmining*, p. 234.
3 *Historical Review*, p. 129.
4 *Ibid*, pp. 150–69.
5 *Report of Charles Morton, H.M.I.M. for the Midlands Inspection District*, 1851, pp. 9–12.
6 *Historical Review*, Appendix, pp. 33–4.
7 Parkinson, *True Stories of Durham Pit Life*, pp. 19–20.
8 *Children's Employment Commission, (Mines) 1842*, App. Pts. i and ii, *passim*.
9 Galloway, *Annals of Coalmining*, pp. 520–1, also *Report of Children's Employment Commission* (Mines) 1842.
10 McCutcheon, *The Hartley Colliery Disaster*.
11 E. A. Martin, *A Piece of Coal*, London, 1896, p. 92.
12 *Historical Review*, pp. 150–1.
13 *Report of Charles Morton*, pp. 8–9.
14 *Ibid*, p. 11.
15 T. H. Hair, *Sketches*, pp. 11 and 12.

CHAPTER SIX *Drainage*

1 Galloway, *Annals of Coalmining*, pp. 46 and 52; they may, of course, have been ditches.
2 Atkinson, *The Great Northern Coalfield*, p. 23.
3 J. U. Nef, *The Rise of the British Coal Industry*, i, p. 59.
4 Quoted, *ibid.*
5 Nef, ii, p. 450.
6 R. S. Smith, 'Huntingdon Beaumont...', *Renaissance and Modern Studies*, i, 1957, pp. 131–4.
7 J. C., *The Compleat Collier*, p. 29.
8 Galloway, *The Steam Engine*, pp. 73–4.
9 A young boy, Humphrey Potter, was said to have introduced self-acting gear, but this story is no more than a pleasant fiction. It was invented by Newcomen himself. See L. T. C. Rolt, *Thomas Newcomen*, David & Charles, 1964. See also J. S. Allen, 'The 1712 and other Newcomen engines of the Earls of Dudley', *Newcomen Soc. Trans.* xxxvii, 1964–5.
10 H. Davey and Others, 'The Newcomen engine', *Proc. Institution of Mechanical Engineers*, Oct.–Dec. 1903.
11 The water is not itself a seal; its purpose is to keep the hemp packing soft.
12 Galloway, *The Steam Engine*, pp. 109–12.
13 *Ibid*, pp. 135–65.
14 *Ibid*, pp. 156, and 191.
15 The Mines Inspector who reported on the Hartley disaster of 1862 said that the accident would not have happened had a Cornish engine been used instead of a Watt engine.
16 I am indebted to Barry Trinder, B.A., Adult Education Tutor with the Shropshire County Council, for bringing these pipes and other artifacts held at Blists Hill to my attention.

CHAPTER SEVEN *Surface Arrangements*

1 J. C., *The Compleat Collier*, pp. 36–8.
2 Galloway, *Annals of Coalmining*, pp. 284–5.
3 Hair, *Sketches*, p. 8.
4 Galloway, *Annals of Coalmining*, p. 484.
5 *Historical Review*, Plate 19.
6 *Ibid*, pp. 208–10.

7 *Ibid*, pp. 211–18.
8 J. C., pp. 34–5.

CHAPTER EIGHT *Coal Face Mechanization*

1 *Historical Review*, pp. 64–76.
2 I. J. Brown, *The Coalbrookdale Coalfield Catalogue of Mines*, Newport, Shropshire, 1968, p. 29.
3 *Historical Review*, pp. 70–2, and J. L. Carvel, *Fifty Years of Machine Mining Progress 1899–1949*, Motherwell.
4 Blundell's collieries are the subject of interesting studies by D. Anderson. See especially 'Blundell's Collieries: technical developments 1776–1966', *Trans. of the Hist. Soc. Lancashire and Cheshire*, vol. 119, 1967.
5 Carvell, p. 32; also *Historical Review*, pp. 77–81.
6 F. J. Anderson and R. H. Thorpe, 'A century of coal-face mechanization', *The Mining Engineer*, no. 83, Aug. 1967.
7 Carvell, pp. 64–5.

CHAPTER NINE *The Miner and the Mining Community*

1 J. R. Leifchild, *Our Coal and Our Coal Pits*, London, 1856, p. 167.
2 Atkinson, *The Great Northern Coalfield*, p. 47; for a picture of sinkers see McCutcheon, *The Hartley Colliery Disaster*, illustration facing p. 49.
3 Parkinson, *True Stories*, pp. 20–1.
4 Cited G. M. Trevelyan, *English Social History* (illustrated edition), Longmans, 1958, iv, p. 137.
5 Children's Employment Commission (Miner), *First Report*, 1842, xv, p. 24.
6 Nef, *The Rise of the British Coal Industry*, ii, p. 188.
7 D. Bremmer, *The Industries of Scotland*, Edinburgh, 1869, p. 5, cited B. F. Duckham, 'Serfdom in eighteenth-century Scotland', *History*, liv, June 1969.
8 Atkinson, p. 44.
9 Leifchild, p. 190.
10 *Coal and Power* (The Report of an Enquiry presided over by the Right Hon. D. Lloyd George, O.M., M.P.) London, n.d., pp. 120–39.

Gazetteer

This gazetteer is necessarily incomplete. Generally, the sites of old mine shafts are not given because they are far too numerous to mention. Some are shown on Ordnance Survey maps; and for students who are particularly interested in them reference may be made to the detailed plans held in the regional or Area offices of the National Coal Board. In most cases there is nothing much to be seen, except maybe a slight depression in the ground. Some old shafts are, however, bricked over or fenced. Marginally more interesting are shafts where the infilling material has so consolidated as to leave a few feet of the shaft lining exposed. Two such shafts are, for example, to be found at the side of a public footpath about four hundred yards from the B6014 road at Whiteborough between Teversal and Tibshelf near the Nottinghamshire–Derbyshire boundary. Similarly, there are two shafts (one of which is octagonal) exposed near Engine Lane, Tyldesley, Lancashire.

Old mine sites are often found near public footpaths many of which were originally for the use of colliers, or were wagonways leading to a canal or coal wharf.

An excellent gazetteer of colliery (and other) wagonways or tramroads is to be found in Bertram Baxter's *Stone Blocks and Iron Rails* (*Tramroads*) published by David & Charles in 1966. It is not proposed to duplicate this information here. Examples of plateways for horse-drawn traffic in fairly recent use are given in an interesting article by I. C. Dodsworth, 'Recent Uses of Horse Plateways in West Yorkshire', *Industrial Archaeology*, vii, May 1970, pp. 131–43.

Museums

Three major open-air projects now actively in preparation are of particular interest to students of the industrial archaeology of coal mining. They are:

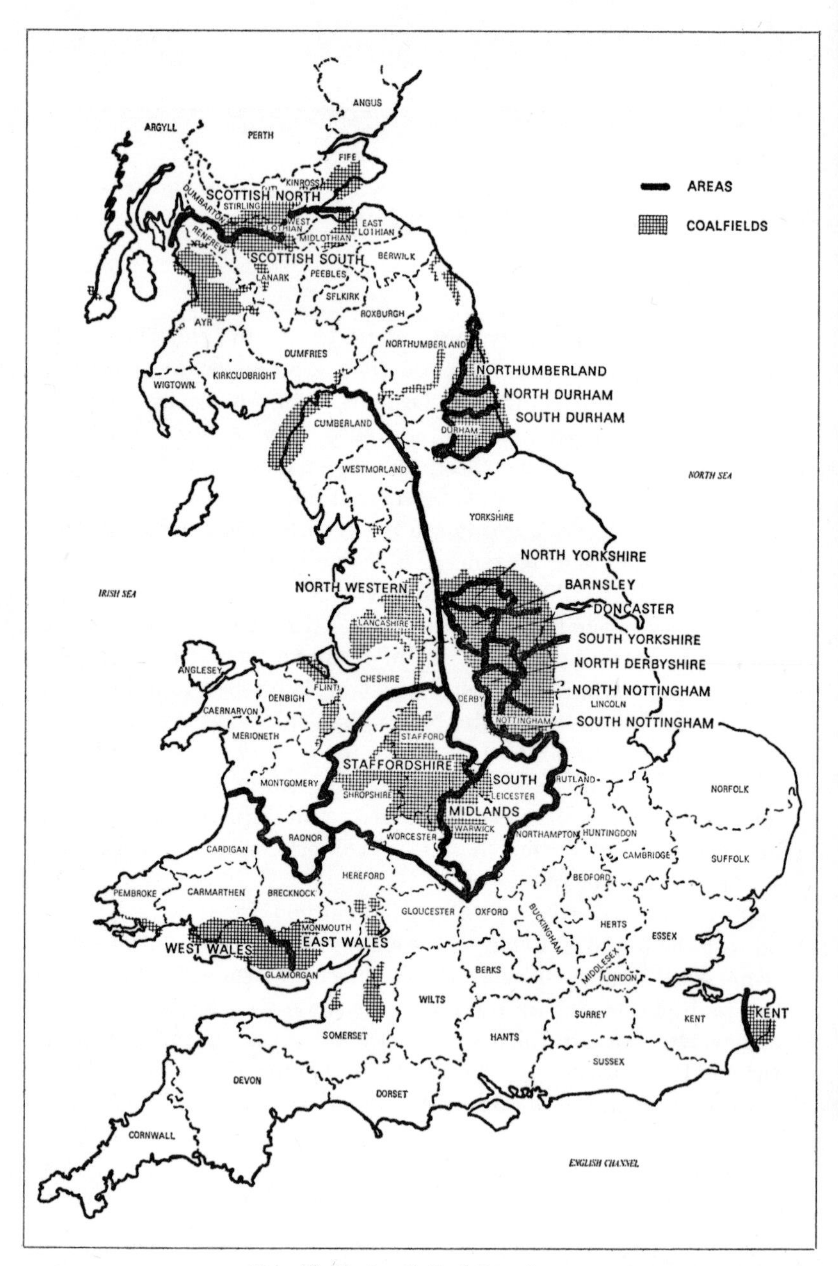

Fig. 18 National Coal Board areas

The Northern Regional Open Air Museum, Beamish, County Durham. Frank Atkinson of the Bowes Museum, Barnard Castle, is in charge of this project and considerable help has been given by the National Coal Board. Items reserved for this Museum include the flat-rope vertical winding engine from Wearmouth; the Beamish Colliery winder; and the Coal Drop from Seaham Harbour illustrated in *Industrial Archaeologists' Guide 1969–70*, published by David & Charles and edited by Neil Cossons and Kenneth Hudson.

Ironbridge Gorge Museum, Blists Hill Site (Shropshire). This is a very worthwhile project which has some support from the authorities but which relies heavily on private individuals, and particularly its own members who do most of the work themselves besides paying their subscriptions.

The Blists Hill site is itself interesting, having remains of a colliery (Shawfield), brickworks, ironworks, canal and tramway all of which have coalmining connections. All aspects of coalmining in Shropshire will be illustrated by the exhibits being acquired. It is intended to rebuild several old headstocks transferred from local collieries. Other items include cages and winding gear; engines; a fan; clay drainage pipes; trams and rails.

Prestongrange Historical Site. This museum is being sponsored by the East Lothian County Council with the help of a committee on which various interests are represented, including the National Coal Board.

At the centre of the site is the old Prestongrange Colliery Beam Engine which drained the workings from 1874 to 1954. This engine was constructed by Harvey & Co. of Hayle, Cornwall in 1874; and in 1895 another firm, the Summerlee Iron Company, increased its capacity to 900 gallons a minute. The beam weighs approximately thirty tons, is thirty-three feet long and six feet four inches deep at the centre. The cylinder is seventy inches in diameter with a twelve foot stroke.

It is intended to collect various artifacts to illustrate the history of coalmining in Scotland. These will be displayed in the old colliery power house.
(Information kindly communicated by J. Jeffreys, Area Chief Engineer, N.C.B., Scottish South Area.)

Lound Hall Mining Museum. The present author has a personal connection with a more modest project, the Mining Museum at the Lound Hall Mining Training Centre, near Retford, Nottinghamshire. This houses a wagon and two stone sleepers from the Little Eaton tramroad; old wooden trams; harness for pit ponies; a 'guss' (human harness) from Somerset; a wooden sledge used to haul corves, and many small items like miners' lamps. At the time of writing preparations are in hand for dismantling the wooden tandem headgear at Brinsley Colliery, near Eastwood and re-erecting it at Lound Hall.

Other museums with mining exhibits (special items in brackets):
Birmingham Museum of Science and Industry.
Buile Hill Park Museum, Near Samlesbury, Lancashire (Model coal mine and various items of equipment).
Bristol City Museum (Winding Engine from Old Mills).
Cardiff; National Museum of Wales.
Cusworth Hall Museum, Cusworth, Doncaster.
Mercer Museum and Art Gallery, Clayton-le-Moors, near Accrington, Lancashire.
Derby Museum and Art Gallery (Separate industrial museum being established).
Royal Scottish Museum, Edinburgh (*Wylam Dilly*, Colliery Loco built 1813).
Shibden Hall Folk Museum, Halifax (Whim-gin).
Leicester Museum and Art Gallery (Separate museum of technology being established).
Science Museum, South Kensington, London (Engines; lamps; models).
Mining Department, Newcastle University (Mining engineering exhibits).
Newcastle upon Tyne Museum of Science and Engineering (Killingworth Colliery Loco, built 1830).
Nottingham Industrial Museum, Wollaton Hall (Whim-gin from Pinxton).
Salford Science Museum.
Shrewsbury Borough Library and Museum (Siskol Electric Coal Cutter; wooden flanged wheel from Caughley).
Shropshire Mining Club Headquarters, Church Aston, Near Newport (Post from whim-gin; hand winches).

Holman Museum, Camborne, Cornwall (Cornish beam engine).
Middleton Railway, Leeds. This is not primarily a Museum, but a privately operated railway some three miles in length with locos running on standard gauge track. However, there is a small museum of exhibits from the old colliery railway which was authorized by Act of Parliament and opened in 1758. The railway was of four feet one inch gauge until steam locos were introduced in 1812 when it was converted to five foot gauge. Rack-rail locos were used until 1835 when the line reverted to horse traction until 1866 (see B. Baxter, *Stone Blocks and Iron Rails*, David & Charles, 1966, p. 32).

THINGS TO SEE

Arranged by Coal Board Areas (see map)

SCOTTISH NORTH AREA

Devon No. 1 Pit
Cornish pump bellcrank. (The Area Chief Engineer, J. D. Blelloch. has also kindly supplied a list of old equipment in existence a few years ago but now regrettably scrapped.)

Brora Colliery, Sutherland
This is a unique highland coalmining community. The mine is now owned by the workmen themselves.

Avonbraes Colliery, Quarter Hamilton
This is a small drift mine (now closed) which employed twenty men underground and five on the surface in 1964.

Culross Colliery, Fife
Bases of protective walls of the Moat Pit to be seen on the foreshore at Culross at low tide.

NORTHUMBERLAND AREA

Ashington Colliery
Koepe Tower Winder installed in 1922 and in use until recently. One of the earliest such installations in Britain. Capel fans, still in use, installed in 1910.

Pegswood Colliery (now closed)
Waddle fan, about seventy-five years old.
This Area also has an old (*c.* 1920) 'Black Star' coalcutter.

NORTH DURHAM AREA

Philadelphia Workshops, Houghton-le-Spring
Braddyll Locomotive, built 1837 by Timothy Hackworth.

Westoe Colliery, South Shields
Overhead trolley loco system installed over sixty years ago.

Whinfield Works, Rowlands Gill
Five bee-hive coke ovens, built 1861, last worked 1958.

SOUTH DURHAM AREA

Early nineteenth-century stone-built miners' cottages at Leasingthorn (Stone Row) and Oakenshaw (Cross Row).

CUMBERLAND

Saltom Colliery
Situated half a mile south of Whitehaven harbour, sunk 1729 and closed 1848, ruined engine house, sea wall and part of gin house.

William Pit, Bransty, Whitehaven
Sunk 1804 and closed in 1955. Substantial remains of engine houses and other surface features.

Duke and Wellington Pits, Whitehaven
Architecturally interesting remains of engine houses and chimneys of quasi-gothic appearance on the south side of Whitehaven harbour.

Haig Colliery, Bransty, Whitehaven
Inclined plane to harbour and other surface features. This is Cumberland's last working colliery.

Jane Pit, Harrington Road, Workington
Sunk in 1843, this is situated to the south-west of the town. Surface features (engine house, boiler house, chimney, etc.) preserved by Workington Borough Council. Until 1969 it was connected to the old workings of King pit (dating from 1795) and the William and Wellington pits.

Howgill, Whitehaven
A roadway in stone, dippping to meet a seam of coal. The bricked entrance is still open.

NORTH WESTERN
(LANCASHIRE, CHESHIRE AND NORTH WALES)

Astley Green Colliery, Tyldesley

Steam winding engines, details as under:

	Bore inches	*Stroke inches*	*Maker*	*Steam Pressure*	*Valve gear*
No. 1 Pit	35 and 60	72	Yates & Thom	160 lb sq. in.	Corliss
No. 2 Pit	35 and 60	72	Yates & Thom	160 lb. sq. in.	Corliss

Bank Hall Colliery

Steam winding engines, details as under:

	Bore inches	*Stroke inches*	*Maker*	*Steam pressure*	*Valve gear*
No. 1 Pit	26	60	Worsley Mesnes	120	Piston type slide valves
No. 4 Pit	38 and 60	72	Yates & Thom	120	Corliss

Sutton Manor Colliery

Steam fan engine; air compressor; steam winding engines with details as under:

	Bore inches	*Stroke inches*	*Maker*	*Steam pressure*	*Valve gear*
No. 1 Pit	28 and 46	60	Fraser & Chalmers	150	Corliss
No. 2 Pit	33 and 55	66	Yates & Thom Worsley Mesnes	150	Corliss

Wood Colliery, Haydock

Steam winding engines with details as under:

	Bore inches	*Stroke inches*	*Maker*	*Steam pressure*	*Valve gear*
No. 2	24	54	Wilkinson, Wigan	90	Slide valve
No. 3	18	48	R. Daglish	90	Slide

			& Co.		valve
No. 4	32	72	Stevenson, Preston	150	Drop valve

Note: Wooden headstocks recently demolished.

Worsley
Remains of underground canal system (now closed). One of the boats is to be exhibited at Lound Hall.

Pewfall Colliery, Ashton-in-Makerfield
Remains of ventilation furnace chimney.

Old Meadows Colliery, Bacup (near A671)
Recently closed drift mine working pillar-and-stall. The iron tubs were hauled up the drift by a chain (see illustration in Owen Ashmore's *Industrial Archaeology of Lancashire*, p. 72). Unfortunately, there is little now to be seen.

Clifton Colliery, near Burnley
Remains of ventilation furnace chimney.

Winter Hill, Horwich
Traces of old pits which were little more than bell pits.

Montcliffe Colliery, Horwich
A small licensed mine. A disused shaft near by still has substantial remains including headstocks, engine houses and tippler.

Castercliffe, between Nelson and Colne
Bell Pits
Note: An old Bolton & Watt beam engine from the Area Workshops has recently been transferred to Manchester University.

NORTH YORKSHIRE AREA

Ackton Hall Colliery
(*a*) Condensing turbo-generator: one of the earliest to be made. Manufactured by C. A. Parsons of Newcastle upon Tyne it was purchased second hand for Ackton Hall Colliery in 1890. The generator gives an output of 32 KW at 110 Volts D.C. It was used to supply lighting to Ackton Colliery until 1922 when it was transferred to Ackworth where it operated until 1953.

(*b*) Bi-pole electric motor, 110 volts manufactured about 1880–90, complete with starter. This motor was driven from the above-mentioned turbo-generator and operated a chopping machine in the colliery stables.

Wakefield
Surface cavities on Kirkhamgate due to collapse of old headings in the Gawthorpe Seam.

Fryston Colliery
A D.C. generator manufactured in 1894 by Wilson Hartnell of Leeds.

Middleton Broom Colliery, Leeds
This is the colliery for which the Middleton Railway was constructed in 1758. It is shown in a famous painting of a collier in *Costumes of Yorkshire*, by G. Walker published in 1814.

The Colliery closed recently and little now remains apart from the railway.

In Middleton woods nearby are many depressions indicative of early bell-pit working.

Stourton, near Leeds (O.S. Sheet No. 233 NE)
Sough draining workings at Rothwell discharging into River Aire. Similar soughs draining into Kippax Beck (O.S. Sheet 259 NW) and canal near Thornhill Combs Colliery (O.S. Sheet 247 NE).

BARNSLEY AREA

Rockley Ironstone Workings (1816)
Remains of engine house for Newcomen engine. Bell pit mounds.

Tankersley Pumping Shaft (adjacent to M1, about four miles south of Barnsley)
Old colliery surface buildings.

Woolley Colliery
Top row—28 stone terraced houses (pre-1871).

Carlton
Five brick-built houses (of the original eighty-one) plus a converted Working Men's Club.

Caphouse Colliery, Overton, Nr Wakefield
Wooden headstocks, steam winding engine.

Flockton, Nr Wakefield

Old railway tunnel, pump engine house belonging to early nineteenth-century colliery.

SOUTH YORKSHIRE AREA

Handsworth Colliery

Steam winding engine made by Daglish in 1903. Has twin cylinders twenty-six inch bore, sixty inch stroke with a sixteen foot diameter parallel drum, wood lagged; and is fitted with a Walker overspeed device.

Until recently it wound coal from the Parkgate Seam (400 yards deep) at sixty winds per hour. It is now disused and may, indeed, have been scrapped by the time this book appears in print.

Wath Main Colliery

Old wooden headstocks, still in use but to be replaced in 1970–1.

Elsecar Old Colliery

Newcomen (atmospheric) engine which pumped water from the workings continuously from 1795 to 1923. It last worked in 1928, and is still maintained.

Fitzwilliam Drainage Scheme

The Elsecar Engine was connected to the Fitzwilliam Drainage Scheme. This Scheme consists of a series of soughs, water levels and pumping stations, and its purpose was to drain the Barnsley Bed workings on the Fitzwilliam Estate. Some of the water levels can still be travelled. The Headquarters of the Scheme (and of all mines drainage in South Yorkshire) is at Westfield House, Rawmarsh. Close by is an interesting engine house which formerly housed a Cornish beam engine, now replaced by an electric pump.

STAFFORDSHIRE

Stafford Colliery, Stoke-on-Trent

Two vertical winding engines and engine houses. Florence sinking engine.

Kemball Training Centre, Stoke-on-Trent

Engine House, stone built and very impressive which formerly housed two vertical winding engines. Now used as training workshop.

Walsall Wood Colliery, South Staffs.
Some surface buildings, of no particular interest, remain. This was, however, the last colliery to use underground furnace ventilation. The colliery closed in 1965, but the furnace went out of use in 1950.

Harecastle Tunnels
These consist of one railway and two canal tunnels, which pass under the westerly extension of the Goldenhill/Kidsgrove ridge. The southern ends are situated in the Chatterley Valley, and the northern ends at Kidsgrove. The railway tunnel is some 2,000 yards long and the canal tunnels, which are each side of the railway tunnel, are 2,850 yards. The tunnels are notable from a mining point of view in that they pass through all the North Staffordshire coal measures from the Half Yards Coal and Ironstone to the Bullhurst Coal Seam. Many of the seams were worked from the canal tunnels, which were also used for water drainage from neighbouring shallow mine workings.

Dudley underground canal
This canal, which carried a great deal of coal, passes under Dudley Castle through old limestone workings.

SHROPSHIRE

Grange Pit, Granville Colliery
Tandem headgear. (This is the only N.C.B. mine now working in Shropshire.)

Brown Clee Hills
Traces of bell pit working.

Blists Hill
Atmospheric engine house (now part of museum); Coalport Tar Tunnel.

Dawley and Horsehay Common
Hand winches (windlasses) and small wooden headgear.

Madeley and Dawley
Old winding chains and flat rope may be found in fields.

Pontesbury
Two winding engine houses now used as dwelling houses.

Milburgh (Broseley)
Horizontal steam winding engine in engine house (disused).

The Pheasant Inn, Broseley
Cellar cut out of coal seam.

NOTTINGHAMSHIRE

Annesley Colliery
Waddle fan.

Bilsthorpe Colliery
Winding engine, horizontal double cylinder, manufactured by Thornewill and Wareham, Burton-on-Trent, in 1886 and used originally as a marine engine; Waddle Fan.

Brinsley Colliery, Eastwood
Tandem headgear (to be removed and re-erected at the Lound Hall Mining Training Centre). Ventilation furnace flue.

Babbington Colliery, Cinderhill (1841–3)
Tandem headgear.

Bentinck Colliery, Kirkby-in-Ashfield
Steam winding engine, originally a marine engine.

DERBYSHIRE

South Normanton
Remains of wooden headstocks and winding engine house of mid-nineteenth century Winterbank pit.

Ireland Colliery
A good example of a nineteenth-century colliery still in production. Vertical winding engine.

Dronfield
Engine house, with a date stone (1871) formerly housing a beam engine; also remains of bee-hive coke ovens, belonging to Summerley Colliery, closed about 1921.

Eckington
'Seldom Seen' pit. Brick-built engine house formerly housing a large beam engine.

Morton
Remains of colliery. An interesting twin horizontal cylinder winding engine from Morton is now waiting re-erection at the Leicester Museum of Technology.

Oakerthorpe, 'Speedwell' pit
Engine house built 1791. The atmospheric engine, by F. Thompson, is now housed at the Science Museum, South Kensington.

Wingerworth
Traces of bell pits.

Shipley estate, Ilkeston (near A6007 road)
Traces of bell pits in wood.

SOUTH MIDLANDS AREA (LEICESTERSHIRE, SOUTH DERBYSHIRE, WARWICKSHIRE)

Speedwell Shaft, Baddesley Ensor, Warwickshire
Pipe work belonging to beam pump (*c.* 1850). The pump itself has recently been dismantled. Plans exhibited at Lound Hall.

Ellistown Colliery, Leicestershire
Fraser Chalmers steam engine (*c.* 1869) bought second-hand from a Lancashire cotton mill driving a two-throw bucket type pump with a piston diameter of eleven inches and three foot stroke. The pump was made by Gimsons of Leicester about 1930.

Cadley Hill Colliery
No. 1 winding engine was manufactured by Thornewill and Wareham of Burton-on-Trent in 1868 and originally used at Bretby Colliery as an endless rope haulage engine. It was converted into a winding engine about fifty years ago.

No. 2 winder was bought from a marine breaker at Barrow-in-Furness and converted into a winding engine by colliery staff. The drums now on the winder are known to date from before 1857. This engine was used in the later stages of the sinking of No. 1 shaft, replacing a donkey engine.

Calcutta Pumping Station, Thringstone, Leicestershire
Engine house erected *c.* 1832 to house a large beam engine since dismantled.

Coleorton, Leicestershire
In a field near Coleorton Hall can be seen the pattern of stall and pillar workings in association with two filled-in shafts. Probably eighteenth-century or earlier.

South Leicester Colliery, Leicestershire
Browett Linley steam fan engine, 70 h.p. and Waddle fan.

Moira, Leicestershire (regarded as part of the South Derbyshire coalfield)
'The Cob', miners' houses, stone built about 1811.

Church Gresley
127–131 Church Street: miners' houses built 1711.

EAST WALES AREA

Cwmbwrgwm Colliery, near Pontypool (approx. one mile west of Abersychan)
Cast iron headframe, formerly used for water balance winding.

Abergorki Colliery
Waddle fan; and steam capstan engine (*c.* 1850).

Elliot Colliery
Horizontal compound steam winding engine.

Crumlin Colliery
Horizontal steam fan engine.

Old Glyn Pits
Vertical single cylinder steam winding engine, with flat rope winding wheel. Beam engine and pump installed 1845.

Marine Colliery
Beam pumping engine (partly dismantled).

Penrikyber
Cast iron beam off an old Cornish engine situated in a large chamber in the No. 2 shaft. Compressed air driven flat rope winding engine, located over an underground staple pit.

Trehafod
Old drift mine (disused) with surface ventilation furnace and chimney.

Blaenavon and Blaina
Early nineteenth-century artisans' dwellings, many built for miners.

SOMERSET (ATTACHED TO N.C.B.'S EAST WALES AREA)

Coalpit Heath Colliery
Remains of colliery (closed 1949); particularly an abandoned haystack boiler.

Parkfield Colliery (1856 to 1936) adjacent to M4 Motorway
Boiler house chimney.

New Rock Colliery, Stratton-on-Fosse (near B2356 road)
One of the three remaining Somerset collieries, employing about 200 men. Has an old four feet six inch diameter shaft and steam winding engine.

Harry Stoke (north of Bristol between A38 and B4058 roads)
Surface remains of modern drift mine which operated only from 1955 to 1963.

Moorwood (near A37, south-west of Stratton-on-the-Fosse)
Substantial surface remains of a nineteenth-century coalmine which closed in 1932.

Fig. 19 Moorwood Colliery, Somerset

Forest of Dean
There are no N.C.B. mines left, but still about twenty small private drift mines which are worth seeing. Most have a small shaft as a second means of egress and to provide ventilation; some of these are operated by hand windlass. The 'free miners' of the Forest of Dean have retained many of their ancient privileges.

The Speech House, where the Mine Law Court was held, is now an hotel. The Verderer's Court still meets there, however. The N.C.B.'s historical film *Nine Centuries of Coal* was made in the Forest of Dean.

The whole area would repay detailed study. Besides providing evidence of mining at various stages from the bell pit era to modern mining, the Forest of Dean exhibits in its drift mines present day survivals of old equipment and methods. The mines currently in production are listed in the *Guide to the Coalfields*, published annually.

DONCASTER AREA

Askern

Steam winding engines.

Yorkshire Main

Typical nineteenth-century mine surface with steam winders, shortly to be replaced by electrically driven ones.

Brodsworth

Waddle fan driven by Bellis & Morcom vertical steam engine. Due to be replaced by a Howden electrically driven fan in 1970.

Hickleton Main

Waddle fan. This is the standby fan and is electrically driven.

Thorne

Steam winders by Fraser & Chalmers. Details as under:

No. 1 Engine Date: 1910
Winding depth: 923 yards
Tandem Compound Type
Cylinders HP 36 inch dia.—Corliss Valves
LP 60 inch dia.—(Inlet Valves Drop Type)
(Exhaust Corliss Type)
Stroke 72 inch
Steam Pressure 150 p.s.i.
Drum: Bicylindro Conical 26 inch dia. to 16 feet dia. 20 feet $5\frac{1}{2}$ inches wide
Blacks Brakes
Whitmore Reversing Engine
Ropes were 2 inch dia. L.C. Type

This is a fine engine of its type.

No. 2 Engine Date: 1910
28 inch dia. cylinders
60 inch stroke
Drum 16 feet by 8 feet 6 inches diameter
Parallel drum, double coiling, winding 863 yards from High Hazel.
Post brakes: Whitmore Brake Engine.

The author wishes to acknowledge valuable assistance given by the undermentioned gentlemen in the compilation of this Gazetteer: R. Storer, N.C.B., North Notts Area; A. J. Williams, Chief Engineer, N.C.B., North Notts Area; N. Watson, Chief Engineer, N.C.B., Northumberland Area; J. D. Blelloch, Chief Engineer, N.C.B., Scottish North Area; G. A. Ross, Chief Engineer, N.C.B., North Durham Area; W. N. Fleming, Estates Manager, N.C.B., South Durham Area; E. Loynes, Chief Engineer, N.C.B., North Western Area; J. B. Smethurst (for Lancashire); J. Smith, Chief Engineer, N.C.B., North Yorkshire Area; K. B. Sutcliffe, Chief Engineer, N.C.B., Barnsley Area; J. M. Bellis, Estates Manager, N.C.B., Barnsley Area; S. Collier, Chief Engineer, N.C.B., South Yorkshire Area; R. Ditchfield, Manager, South Yorkshire Mines Drainage Unit; C. Buchan, Chief Engineer, N.C.B., Staffordshire Area; R. H. Howson, Industrial Relations Officer, N.C.B., Staffordshire Area; E. R. Hassall, Surveyor & Minerals Manager, Staffordshire Area, N.C.B.; I. J. Brown (for Shropshire, mainly); B. Trinder (for Shropshire); S. C. Walker, Chief Engineer, N.C.B., South Midlands Area; E. A. Dixon, Estates Manager, N.C.B., South Midlands Area; W. Humphreys, Chief Engineer, N.C.B., East Wales Area; E. J. Murphy, Estates Manager, N.C.B., East Wales Area; Robin Atthill (for Somerset); N. Smedley, Chief Surveyor & Minerals Manager, N.C.B., North Derbyshire Area; V. Bown, Chief Surveyor & Minerals Manager, N.C.B., North Yorkshire Area; W. S. Askew, Chief Engineer, N.C.B., Doncaster Area; C. Humphreys of Whitehaven for information on Cumberland.

Select Bibliography

GENERAL

Ashton, T. S. and Sykes, J. *The Coal Industry of the Eighteenth Century*, Manchester University Press, 1929.

Baxter, B. N. *Stone Blocks and Iron Rails*, David & Charles, 1966.

Cossons, N. and Hudson, K., eds., Industrial Archaeologist's Guide, 1969–70. David & Charles, 1969.

J. C. *The Compleat Collier* (1708), Newcastle upon Tyne, F. Graham, Northern reprints No. 4, 1968.

Gale, W. K. V. *Iron and Steel*, Industrial Archaeology Series, Longmans, 1969.

Galloway, R. L. *Annals of Coalmining*, Macmillan. 2 vols, 1898, 1904.

Galloway, R. L. *The Steam Engine and Its Inventors*, Macmillan, 1881.

Morgan B. *Civil Engineering: Railways*, Industrial Archaeology Series, Longman, 1971.

Moss, K. N. and Others. *Historical Review of Coal Mining*, The Mining Association, 1925.

Nef, J. U. *The Rise of the British Coal Industry*, Routledge, 1932, 2 vols.

Pannell, J. P. M. *Techniques of Industrial Archaeology*, David & Charles, 1966.

Rolt, L. T. C. *Navigable Waterways*, Industrial Archaeology Series, Longman, 1969.

Snell, J. B. *Mechanical Engineering: Railways*, Industrial Archaeology Series, Longman, 1971.

REGIONAL AND LOCAL STUDIES

Anderson, D. 'Blundell's Collieries: technical developments, 1766–1966', *Trans. Historic. Soc. of Lancashire and Cheshire*, vol. 119, 1967.

Atkinson, F. *The Great Northern Coalfield*, 1700–1900, University Tutorial Press, 1968.

Banks, A. G. and Schofield, R. B. *Brindley at Wet Earth Colliery*, David & Charles, 1968.

Broadhead, I. E. 'Signs of the mines', *Colliery Guardian*, 29 Sept. 1967.

Brown, I. J. 'The mineral wealth of Coalbrookdale', reprinted from the *Bulletin of the Peak District Mines Historical Society*, 1965.

Brown, I. J. *The Coalbrookdale Coalfield Catalogue of Mines*, Shropshire County Library, 1968.

Clayton, A. K. 'The Newcomen type engine at Elsecar, West Riding', *Trans. Newcomen Society*, vol. 35, 1962–3, 97–108.

Duckham, Baron F. *A History of the Scottish Coal Industry*, vol. 1, 1700–1815, David & Charles, 1970.

George, A. D. 'The Industrial Archaeology of West Cumberland', *Industrial Archaeology*, Vol. 7, No. 2, May 1970.

Griffin, A. R. *Mining in the East Midlands 1550–1947*, Frank Cass and Co., 1971.

Griffin, A. R. 'Bell pits and soughs: some East Midlands examples', *Industrial Archaeology*, Vol. 6, No. 4, Nov. 1969.

Griffin, A. R. 'On how to write a history of the East Midlands Coal Industry', *Bulletin of Local History—East Midland Region*, 1970.

Hassall, E. R. and Trickett, J. P. 'The Duke of Bridgewater's underground canals', *Trans. Manchester Geological and Mining Society*, 15 Nov. 1962.

Robertson, T. (*Ed.*), *A Pitman's Notebook, The Diary of Edward Smith, Houghton Colliery Viewer, 1749–1751*, Newcastle-upon-Tyne, F. Graham, Northern History Booklets, 1970.

Tait, H. 'The Forgotten art of cannel carving', *Colliery Guardian*, 31 Dec. 1965.

'Industrial Housing: East Shropshire', *Shropshire News Letter*, no. 33, Dec. 1967.

Industrial Archaeology of the British Isles Series, David & Charles:

Ashmore, Owen. *Industrial Archaeology of Lancashire*, 1969.

Buchanan, R. A. and Cossons, N. *The Bristol Region*, 1969.

Butt, John. *Industrial Archaeology of Scotland*, 1967.

Davies-Shiel, M. and Marshall, J. D. *Industrial Archaeology of the Lake Counties*, 1969.

Nixon, Frank. *Industrial Archaeology of Derbyshire*, 1969.

Smith, D. M. *Industrial Archaeology of the East Midlands*, 1965.

Index